McGraw-Hill's

500 MCAT General Chemistry Questions to Know by Test Day

Also in McGraw-Hill's 500 Questions Series

McGraw-Hill's
500 MCAT General Chemistry Questions to Know by Test Day

John T. Moore, EdD
Richard H. Langley, PhD

New York Chicago San Francisco Lisbon London Madrid Mexico City
Milan New Delhi San Juan Seoul Singapore Sydney Toronto

The **McGraw·Hill** *Companies*

John T. Moore, EdD, has taught chemistry at Stephen F. Austin State University for more than 40 years, where he is Director of the Teaching Excellence Center and Codirector of the Science, Technology, Engineering, and Mathematics Center.

Richard H. Langley, PhD, has taught chemistry at the university level for more than 30 years and has coauthored numerous books on the subject. He has written questions for the AP Chemistry exam and taught General Chemistry and Organic Chemistry MCAT review courses.

1 2 3 4 5 6 7 8 9 10 11 12 13 14 15   QFR/QFR   1 9 8 7 6 5 4 3 2

ISBN   978-0-07-178311-8
MHID   0-07-178311-3

e-ISBN   978-0-07-178312-5
e-MHID   0-07-178312-1

Library of Congress Control Number 2011944706

MCAT is a registered trademark of the Association of American Medical Colleges, which was not involved in the production of, and does not endorse, this product.

McGraw-Hill products are available at special quantity discounts to use as premiums and sales promotions or for use in corporate training programs. To contact a representative, please e-mail us at bulksales@mcgraw-hill.com.

This book is printed on acid-free paper.

CONTENTS

INTRODUCTION

Congratulations! You've taken a big step toward MCAT success by purchasing *McGraw-Hill's 500 MCAT General Chemistry Questions to Know by Test Day*. We are here to help you take the next step and score high on your MCAT exam so you can get into the medical school of your choice!

This book gives you 500 MCAT-style multiple-choice questions that cover all the most essential course material. Each question is clearly explained in the answer key. The questions will give you valuable independent practice to supplement your regular textbook and the ground you have already covered in your general chemistry class.

This book and the others in the series were written by expert teachers who know the MCAT inside and out and can identify crucial information as well as the kinds of questions that are most likely to appear on the exam.

You might be the kind of student who needs to study extra a few weeks before the exam for a final review. Or you might be the kind of student who puts off preparing until the last minute before the exam. No matter what your preparation style, you will benefit from reviewing these 500 questions, which closely parallel the content, format, and degree of difficulty of the questions on the actual MCAT exam. These questions and the explanations in the answer key are the ideal last-minute study tool for those final weeks before the test.

If you practice with all the questions and answers in this book, we are certain you will build the skills and confidence needed to excel on the MCAT. Good luck!

—Editors of McGraw-Hill Education

McGraw-Hill's

500 MCAT
General Chemistry
Questions
to Know by Test Day

CHAPTER 1

The Basics

1. How many grams of hydrogen are in 5 moles of phosphoric acid?
   - (A) 100 g
   - (B) 3.0 g
   - (C) 490 g
   - (D) 15 g

2. Determine the percent mass of each element in H_2SO_4.
   - (A) 29% hydrogen, 14% sulfur, 57% oxygen
   - (B) 2.0% hydrogen, 33% sulfur, 65% oxygen
   - (C) 4.0% hydrogen, 32% sulfur, 64% oxygen
   - (D) 2.0% hydrogen, 48% sulfur, 50% oxygen

3. Copper(II) nitrate decomposes when heated according to the following equation:

$$2\,Cu(NO_3)_2 \rightarrow 2\,CuO + 4\,NO_2 + O_2$$

   How many grams of NO_2 form when 2.0 moles of copper(II) nitrate decompose?
   - (A) 46 g
   - (B) 92 g
   - (C) 180 g
   - (D) 220 g

4. What is the oxidation state of iron in potassium ferrate, K_2FeO_4?
   - (A) +6
   - (B) +4
   - (C) +3
   - (D) +2

5. Which of the following contains an element in a +1 oxidation state?

   (A) Cu_2S
   (B) CuS
   (C) $Ca(ClO_2)_2$
   (D) $AlCl_3$

6. The formula of pyruvic acid is $HC_3H_3O_3$. What is the mass percent of oxygen in pyruvic acid?

   (A) 30%
   (B) 55%
   (C) 41%
   (D) 60%

7. The following equation illustrates the formation of heme from ferrous ion and protoporphyrin IX:

   $$C_{31}H_{28}N_4O_4(aq) + Fe^{2+}(aq) \rightarrow FeC_{31}H_{26}N_4O_4(aq) + 2\,H^+(aq)$$

   What type of reaction is the formation of heme?

   (A) Lewis acid-base
   (B) Oxidation-reduction
   (C) Single replacement
   (D) Neutralization

8. The formation of window glass involves the fusing of a mixture of sand, SiO_2; calcium carbonate, $CaCO_3$; and sodium carbonate, Na_2CO_3. The process generates a gas. What is the identity of the gas?

   (A) CO_2
   (B) O_2
   (C) Na_2O
   (D) O_3

9. What is the concentration of ammonium ions in a 0.3 M solution of ammonium phosphate?

   (A) 0.6 M
   (B) 0.3 M
   (C) 0.9 M
   (D) 0.0 M

10. How many milliliters of 0.40 *M* NaOH solution are necessary to neutralize 0.0500 L of 0.20 *M* HCl solution?
    (A) 0.025 mL
    (B) 50.0 mL
    (C) 75.0 mL
    (D) 25.0 mL

11. The mass percent of oxygen in sucrose, $C_{12}H_{22}O_{11}$, is approximately
    (A) 50%
    (B) 25%
    (C) 10%
    (D) 75%

12. The density of water is 1 g/mL. What is the molarity of water in pure water?
    (A) 55 *M*
    (B) 5.5 *M*
    (C) 18 *M*
    (D) 1 *M*

13. What is the oxidation state of carbon in formaldehyde (HCHO)?
    (A) 0
    (B) +4
    (C) −4
    (D) +2

14. What is the oxidation state of iodine in the periodate ion (IO_4^-)?
    (A) 0
    (B) +8
    (C) −1
    (D) +7

15. The following reaction illustrates the ability of hydrogen to serve as a reducing agent:

$$Re_2O_7(s) + 7\ H_2(g) \rightarrow 2\ Re(s) + 7\ H_2O(g)$$

A 1.0 kg sample of $Re_2O_7(s)$ is reacted with 1.0 kg of H_2. What is the maximum amount of Re formed?

(A) 0.77 kg
(B) 2.0 kg
(C) 0.50 kg
(D) 0.25 kg

16. The solution formed by mixing 100 mL of 3 M H_2SO_4 with 200 mL of 3 M KOH is identical to 300 mL of

(A) 2 M K_2SO_4
(B) 3 M K_2SO_4
(C) 1 M K_2SO_4
(D) 1 M H_2SO_4 plus 2 M KOH

17. Dichlorine heptoxide, Cl_2O_7, is an unstable compound. Which of the following substances contains chlorine in the same oxidation state as the chlorine in dichlorine heptoxide?

(A) HCl
(B) $HClO_4$
(C) $HClO_3$
(D) Cl_2

18. What is the oxidation state of iron in Fe_3O_4?

(A) +8/3
(B) +3
(C) +2
(D) 0

19. Glucose, or blood sugar, is a carbohydrate. The molecular formula for glucose is $C_6H_{12}O_6$. What is the empirical formula of glucose?

(A) CHO
(B) $C_6H_{12}O_6$
(C) CH_2O
(D) $C_3H_4O_2$

20. How many milliliters of 0.20 M $Ba(OH)_2$ solution are necessary to neutralize 0.0500 L of 0.20 M HCl solution?

(A) 75 mL
(B) 50 mL
(C) 12 mL
(D) 25 mL

21. What is the concentration of hydrogen ions in a solution made by mixing 250.0 mL of 0.10 M NaOH with 250.0 mL of 0.20 M HCl?

(A) 0.050 M
(B) 0.10 M
(C) 0.010 M
(D) 0.20 M

22. A sample contains 3.01×10^{23} of C_2H_5OH molecules. How much does the sample weigh?

(A) 46 g
(B) 23 g
(C) 12 g
(D) 92 g

23. What is the molality of a solution containing 16 g of CH_3OH in 500.0 g of water?

(A) 2.0 m
(B) 0.50 m
(C) 1.0 m
(D) 16 m

24. An aqueous mixture of H_2SO_4, $KMnO_4$, and $H_2C_2O_4$ reacts as follows:

$$3 \ H_2SO_4(aq) + 2 \ KMnO_4(aq) + 5 \ H_2C_2O_4(aq)$$
$$\rightarrow K_2SO_4(aq) + 2 \ MnSO_4(aq) + 10 \ CO_2(g) + 8 \ H_2O(l)$$

In one experiment, the initial mixture contained 4 moles of H_2SO_4, 2 moles of $KMnO_4$, and 5 moles of $H_2C_2O_4$. Approximately how many grams of sulfuric acid remain after the reaction?

(A) 100 g
(B) 200 g
(C) 300 g
(D) 150 g

25. The oxidation states of the atoms in formaldehyde, CH_2O, are
   (A) H = 0, C = +2, and O = −2
   (B) H = +1, C = 0, and O = −2
   (C) H = +1, C = −2, and O = +2
   (D) H = −1, C = +4, and O = −2

26. Approximately how many grams of phosphorus are in 1.5 moles of P_4?
   (A) 31 g
   (B) 130 g
   (C) 180 g
   (D) 15 g

27. What is the empirical formula of a compound containing 78% silver, 7.5% phosphorus, and 15% oxygen?
   (A) Ag_3PO_4
   (B) Ag_3PO_3
   (C) $AgPO_3$
   (D) $AgPO_4$

28. Determine the oxidation state of chlorine in each of the following compounds: $HClO_2$ and $HClO_4$.
   (A) The oxidation state of chlorine in $HClO_2$ is +3, and in $HClO_4$ it is +7.
   (B) The oxidation state of chlorine in $HClO_2$ is −1, and in $HClO_4$ it is −1.
   (C) The oxidation state of chlorine in $HClO_2$ is −1, and in $HClO_4$ it is +1.
   (D) The oxidation state of chlorine in $HClO_2$ is +3, and in $HClO_4$ it is −1.

29. Ag_2O decomposes when heated according to the following equation:

$$2\ Ag_2O \rightarrow 4\ Ag + O_2$$

How many moles of oxygen are formed by the decomposition of 3.0 moles of Ag_2O?
   (A) 2.0 mol
   (B) 1.0 mol
   (C) 3.0 mol
   (D) 1.5 mol

30. Which of the following is the percent composition of ethyl alcohol (C_2H_5OH)?
    (A) 52% carbon, 11% hydrogen, 37% oxygen
    (B) 52% carbon, 13% hydrogen, 35% oxygen
    (C) 26% carbon, 13% hydrogen, 61% oxygen
    (D) 19% carbon, 11% hydrogen, 70% oxygen

31. Light induces the following reaction in some photographic paper:

$$AgI(s) \rightarrow Ag(s) + I^-(aq)$$

    What type of process is this?
    (A) Ionization reaction
    (B) Oxidation reaction
    (C) Lewis acid-base reaction
    (D) Reduction reaction

32. The formal charges of the atoms in the cyanide ion, CN^-, are
    (A) -1 for carbon and 0 for nitrogen
    (B) 0 for carbon and -1 for nitrogen
    (C) $-\frac{1}{2}$ for each atom
    (D) $+2$ for carbon and -3 for nitrogen

33. The following equation illustrates the formation of heme from ferrous ion and protoporphyrin IX:

$$C_{31}H_{28}N_4O_4(aq) + Fe^{2+}(aq) \rightarrow FeC_{31}H_{26}N_4O_4(aq) + 2\,H^+(aq)$$

    If the final pH of the solution was 5, what was the original pH of the solution?
    (A) Equal to 5
    (B) Less than 5
    (C) Greater than 5
    (D) Cannot be determined

34. Determine the molar concentration of a solution containing 0.20 moles of solute in 100.0 cm³ of solution.
    (A) 20 M
    (B) 0.2 M
    (C) 2.0 M
    (D) 0.02 M

35. Determine the molality of a solution containing 5 moles of ethanol (C_2H_5OH) in 1,000 g of water.

    (A) 0.05 m
    (B) 0.005 m
    (C) 5000 m
    (D) 5 m

36. A solution was prepared by mixing 50.0 mL of 0.10 M $Na_2CO_3(aq)$ with 50.0 mL of 0.10 M $NaCl(aq)$. What is the concentration of $Na^+(aq)$ in this solution?

    (A) 0.15 M
    (B) 0.10 M
    (C) 0.050 M
    (D) 0.075 M

37. A sample contains Avogadro's number of water molecules. How much does the sample weigh?

    (A) 6 × 10^{23} g
    (B) 18 g
    (C) 1 g
    (D) 17 g

38. In the following reaction, how many moles of H_2S form when 2.0 moles of HCl react with 0.5 moles of FeS?

$$2\ HCl(g) + FeS(s) \rightarrow FeCl_2(s) + H_2S(g)$$

    (A) 2.5 mol
    (B) 2.0 mol
    (C) 0.5 mol
    (D) 1.0 mol

CHAPTER 2

Atomic Structure

39. An isotope of calcium undergoes electron capture and becomes an isotope of which element?

(A) Calcium
(B) Scandium
(C) Potassium
(D) Argon

40. The actual mass of an isotope is less than the sum of the masses of the constituents by an amount known as the

(A) atomic weight
(B) atomic mass
(C) mass defect
(D) molar mass

41. Which element might have the following excited-state electron configuration: $1s^2 2s^2 2p^5 3s^2 3p^1$?

(A) Al
(B) Mg
(C) Na
(D) Ar

42. Which is the only noble gas that does NOT have an ns^2np^6 outer shell electron configuration?

(A) He
(B) Ne
(C) Ar
(D) Rn

43. If an isotope of gold undergoes alpha decay, an isotope of which element will result?

    (A) Hg
    (B) Pt
    (C) Ir
    (D) Tl

44. Which element has the following electron configuration: $[Kr]5s^24d^7$?

    (A) Ir
    (B) Co
    (C) Rh
    (D) Tc

45. Copper-57 undergoes positron decay with a half-life of 200 ms. Sodium-30 undergoes beta decay with a half-life of 50 ms. How does the rate constant for the decay of sodium-30 compare with the rate constant for the decay of copper-57?

    (A) Smaller by a factor other than 4
    (B) Smaller by a factor of 4
    (C) Larger by a factor other than 4
    (D) Larger by a factor of 4

46. What radioactive decay mode is involved in the conversion of ^{86}Br to ^{86}Kr?

    (A) β^+ decay
    (B) β^- decay
    (C) α decay
    (D) Electron capture

47. The isotope ^{100}Pd undergoes electron capture. Which isotope does this process produce?

    (A) ^{100}Ag
    (B) ^{100}Rh
    (C) ^{96}Ru
    (D) ^{99}Pd

48. The radioisotope iodine-131 is useful in studying the function of the thyroid gland. The isotope has a half-life of 8.0 days. What percentage of the iodine-131 remains in a sample after 40 days?

(A) 12.5%
(B) 6.25%
(C) 3.13%
(D) 50.0%

49. An isotope of an element undergoes a series of radioactive decay processes to produce a different isotope of the same element. If the first step was an alpha decay, how many beta decay steps were there?

(A) four
(B) one
(C) zero
(D) two

50. Electric fields affect all nuclear decay products EXCEPT

(A) α particles
(B) γ rays
(C) β^+ particles
(D) β^- particles

51. A radioisotope undergoes decay with a half-life of 8 minutes. How long will it take for only 5% of the sample to remain?

(A) 16 minutes
(B) 35 minutes
(C) 65 minutes
(D) 8 minutes

52. Which of the following compounds has nitrogen in the highest oxidation state?

(A) HNO_3
(B) NH_3
(C) N_2
(D) $H_2N_2O_2$

53. The electron configuration for a possible excited state of the Ag^+ ion would be
    (A) $[Kr]4d^{10}5s^1$
    (B) $[Kr]4d^{10}$
    (C) $[Kr]4d^95s^1$
    (D) $[Kr]4d^9$

54. Tritium, ^3H, is a beta emitter. What isotope does the beta emission of tritium produce?
    (A) ^4He
    (B) ^2H
    (C) ^1H
    (D) ^3He

55. If radium-228 undergoes two beta decays and an alpha decay, what isotope forms?
    (A) Radium-224
    (B) Thorium-228
    (C) Thorium-224
    (D) Actinium-228

56. The n = 4 shell can hold a maximum of
    (A) 2 electrons
    (B) 8 electrons
    (C) 18 electrons
    (D) 32 electrons

57. One of the electrons on a gold atom has a quantum number of -3. Which subshell holds this electron?
    (A) 6s
    (B) 5d
    (C) 4f
    (D) 5p

58. The half-life of carbon-14 is about 5,700 years. Approximately how long will it take for 1 g of carbon-14 to decay to 0.23 g?
    (A) 5,700 years
    (B) 11,400 years
    (C) 17,100 years
    (D) 22,800 years

59. Analysis of the air in a container with an unknown radioactive isotope has a high concentration of helium gas. What type of radioactive decay does the unknown radioisotope experience?

(A) Positron emission
(B) Beta emission
(C) Gamma decay
(D) Alpha emission

60. The p orbitals on a nitrogen atom are degenerate. What does *degenerate* mean?

(A) The energies of the orbitals are the same.
(B) There are three orbitals in the p orbital set.
(C) The orbitals can form pi (π) bonds.
(D) The orbitals can hybridize.

61. Iodine-131 is a beta emitter. The nuclear accident at Fukushima, Japan, released quantities of radioactive iodine-131 into the atmosphere.
Iodine is an essential element for humans. The product of the radioactive decay of iodine-131 is

(A) ^{131}Xe
(B) ^{130}I
(C) ^{127}Sb
(D) ^{131}Te

62. Which of the following qualifies as a Bohr atom?

(A) He^+
(B) H^-
(C) H^+
(D) Li^+

63. Certain numbers of protons or neutrons impart additional stability to a nucleus. These numbers are the "magic numbers" (2, 8, 20, 28, 50, 82, 126 . . .). Nuclei with a magic number of either protons or neutrons are less likely to be radioactive (doubly magic are even more stable). Which of the following nuclei is LEAST likely to be radioactive?

(A) ^8Li
(B) ^{12}B
(C) ^8Be
(D) ^{15}N

64. Certain numbers of protons or neutrons impart additional stability to a nucleus. These numbers are the "magic numbers" (2, 8, 20, 28, 50, 82, 126 . . .). Nuclei with a magic number of either protons or neutrons are less likely to be radioactive (doubly magic are even more stable). Which of the following is a doubly magic nucleus?

    (A) ^{57}Fe
    (B) ^{118}Sn
    (C) ^{208}Pb
    (D) ^{17}O

65. Stable light isotopes normally have approximately equal numbers of protons and neutrons. Stable heavier isotopes normally have more neutrons than protons. Isodiapheres are isotopes of different elements with the same number of excess neutrons over protons. Which of the following pairs represent isodiapheres?

    (A) ^{238}U and ^{234}Th
    (B) ^{208}Pb and ^{202}Hg
    (C) ^{128}Xe and ^{130}Ba
    (D) ^{16}O and ^{22}Ne

66. Which of the following radioactive nuclei is LEAST likely to be harmful in contact with skin?

    (A) ^{187m}Pb: half-life = 15.2 s, alpha emission; particle energy = 5.99 MeV
    (B) ^{230}Ac: half-life = 2.0 m, beta emission; particle energy = 1.4 MeV
    (C) ^{211}At: half-life = 7.21 h, electron capture; particle energy = 0 MeV
    (D) ^{232}Th: half-life = 1.4 $\times$ 10^{10} y, alpha emission; particle energy = 2.83 MeV

67. A laboratory accident resulted in the release of radioactive radon222 ($t_{1/2}$ = 3.8 days). The initial radiation level was 1×10^{-5} μCi/mL (microcuries per milliliter). Even with protective respiratory gear, it is not safe to enter the laboratory until the radiation level drops below 1×10^{-6} μCi/mL. What is the LEAST number of days required for the radiation level to drop below 1×10^{-6} μCi/mL?

    (A) 3.8 days
    (B) 10 days
    (C) 12 days
    (D) 16 days

68. Iron-64 is radioactive with a half-life of 2.0 s and emits beta particles. The BEST explanation of why iron-64 is a beta emitter is that

(A) the neutron-electron ratio is too high
(B) the neutron-proton ratio is too low
(C) the neutron-proton ratio is not 1
(D) the neutron-proton ratio is too high

69. Which of the following is most likely to occur during an electron transition from the 3p to the 1s subshell?

(A) Energy is emitted, and there is a bright band in the atomic spectrum.
(B) Energy is emitted, and there is a dark band in the atomic spectrum.
(C) Energy is absorbed, and there is a bright band in the atomic spectrum.
(D) Energy is absorbed, and there is a dark band in the atomic spectrum.

70. If a radioactive atom emits two alpha particles, by how many units does its mass change?

(A) 8
(B) 4
(C) 2
(D) 0

71. How many orbitals are in the 7d subshell?

(A) 5
(B) 7
(C) 3
(D) 1

72. Iron may form an Fe^{3+} ion. What is the ground-state electron configuration of Fe^{3+}?

(A) $[Ar]4s^23d^6$
(B) $[Ar]4s^23d^3$
(C) $[Ar]$
(D) $[Ar]3d^5$

73. The half-life of hydrogen-3 is about 14 years. What is the approximate amount remaining after 8 years?

(A) 67%
(B) 50%
(C) 90%
(D) 25%

74. Based upon the electron configurations listed, which of the following ions is NOT paramagnetic?
    (A) Ca^+ $[Ar]4s^1$
    (B) Fe^{3+} $[Ar]3d^5$
    (C) Cu^+ $[Ar]3d^{10}$
    (D) Ga^{2+} $[Ar]3d^{10}4s^1$

75. Which is a representation of the electron configuration of silicon in an excited state?
    (A) $1s^22s^22p^63s^23p^14s^1$
    (B) $1s^22s^22p^63s^23p^24s^1$
    (C) $1s^22s^22p^63s^23p^14s^2$
    (D) $1s^22s^22p^63s^23p^3$

76. The radioactive decay series of ^{232}Th ends with ^{208}Pb. How many α and β particles are emitted during this decay?
    (A) 6 α and 7 β particles
    (B) 7 α and 7 β particles
    (C) 8 α and 6 β particles
    (D) 7 α and 6 β particles

77. A nuclear fusion process converted 2.0×10^{-28} g to energy. How much energy was released?
    (A) 1.8×10^{-11} J
    (B) 1.8×10^{-14} J
    (C) 6.0×10^{-20} J
    (D) 6.0×10^{23} J

78. If two nuclei fuse to produce another isotope and release energy, what can be said about the exact mass of the nucleus formed?
    (A) It is equal to the sum of the two fused nuclei.
    (B) It is greater than the sum of the two fused nuclei.
    (C) It is less than the sum of the two fused nuclei.
    (D) It is equal to the average of the two fused nuclei.

79. A valence electron on a sodium atom might have which of the following sets of quantum numbers?

   (A) 3, 1, 0, −½
   (B) 3, 0, 1, +½
   (C) 2, 0, 0, +½
   (D) 3, 0, 0, +½

80. An electron on an atom has the following quantum numbers: 5, 2, 0, +½. How many other electrons on the same atom can have this set of quantum numbers?

   (A) 3
   (B) 2
   (C) 1
   (D) 0

81. How many sp² hybridized carbon atoms are in the following molecule?

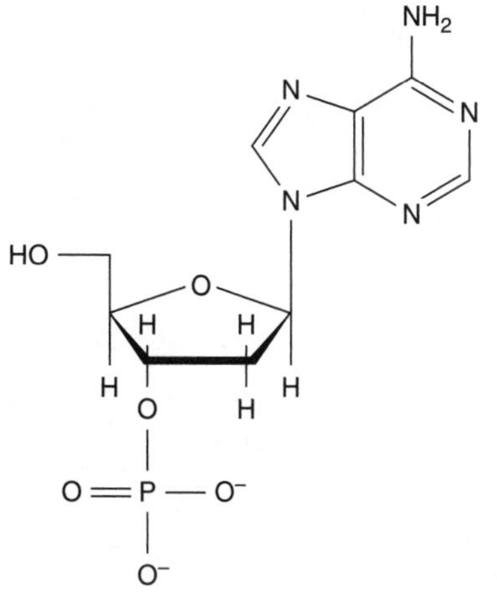

   (A) 9
   (B) 5
   (C) 15
   (D) 3

82. Which of the following transition metal compounds exhibits color?

    (A) ZnO
    (B) TiO_2
    (C) Fe_2O_3
    (D) Sc_2O_3

83. A nuclear accident released quantities of radioactive iodine-131 into the atmosphere. Iodine is an essential element for humans. The concern at the time was that humans would absorb the radioactive iodine along with natural (nonradioactive) iodine. One treatment to reduce the amount of iodine-131 absorbed would be to have potential victims

    (A) ingest diuretics to maximize water (containing iodine-131) loss
    (B) boil possibly contaminated food and water to increase the radioactive decay of iodine-131
    (C) ingest silver nitrate tablets to precipitate the iodine as silver iodide
    (D) ingest large amounts of nonradioactive iodine

CHAPTER 3

Bonding and the Periodic Table

84. How does the atomic radius of potassium relate to the atomic radius of arsenic?
  (A) Potassium is smaller because the effective nuclear charge is smaller.
  (B) Potassium is larger because the effective nuclear charge is larger.
  (C) Potassium is larger because the effective nuclear charge is smaller.
  (D) Potassium is smaller because the effective nuclear charge is larger.

85. In the Lewis structure of $MgCl_2$, how many electrons are shared?
  (A) 0
  (B) 1
  (C) 2
  (D) 8

86. How does the third ionization energy of Ca compare with the third ionization energy of Sc?
  (A) The value for Ca is much higher.
  (B) The value for Ca is much lower.
  (C) The values are the same.
  (D) Unrelated because Sc is a transition element and Ca is not.

87. Which of the following is the correct Lewis structure for O_2?
  (A) $\ddot{O} = \ddot{O}$
  (B) $\ddot{O} - \ddot{O}$
  (C) $:O = O:$
  (D) $:\ddot{O} - \ddot{O}:$

88. An excited-state potassium atom emits light of characteristic wavelengths. These emitted wavelengths are due to

(A) excited-state electrons moving to higher energy levels
(B) excited-state electrons dropping to lower energy levels
(C) transitions within the potassium nucleus
(D) excited-state potassium atoms interacting with ground-state potassium atoms

89. Ethyl alcohol has a normal boiling point of 78°C. The normal boiling point is determined at

(A) a pressure of 760 torr
(B) a pressure below 760 torr
(C) a pressure above 760 torr
(D) all pressures

90. The radius of a nitrogen atom is about 75 pm. The addition of electrons to a nitrogen atom results in an increase in the radius. Each added electron increases the radius by about 20%. What is the approximate radius of a nitride ion?

(A) 138 pm
(B) 110 pm
(C) 90 pm
(D) 60 pm

91. The rate of the passive diffusion of ions across a cell membrane is inversely proportional to the radius of the ion. Which of the following ions has the slowest rate of passive diffusion?

(A) F^-
(B) K^+
(C) Ba^{2+}
(D) Mg^{2+}

92. Many properties of the elements show a general periodic trend. What is the general trend in the ionization energies of the elements?

(A) Increases to the left and downward
(B) Decreases to the left and downward
(C) Decreases to the left and increases downward
(D) Increases to the left and decreases downward

93. Why is the second electron affinity of an atom significantly greater than the first electron affinity?

(A) This makes an electron capture process more likely.
(B) This completes the stable electron configuration.
(C) The second electron is held more tightly than the first.
(D) There is repulsion of like charges.

94. Why is the ionization energy of sodium lower than that of chlorine?

(A) The valence electron of sodium is closer to the nucleus and experiences a lower effective nuclear charge.
(B) The valence electron of sodium is farther from the nucleus and experiences a lower effective nuclear charge.
(C) The valence electron of sodium is farther from the nucleus and experiences a higher effective nuclear charge.
(D) The valence electron of sodium is closer to the nucleus and experiences a higher effective nuclear charge.

95. Oxygen molecules will bind to the iron in hemoglobin. When this occurs, the oxygen molecule is acting as

(A) an oxidizing agent
(B) a Lewis acid
(C) a Lewis base
(D) a reducing agent

96. The BEST way of describing the binding of oxygen to the iron in hemoglobin is the formation of

(A) a dipole-dipole interaction
(B) an ionic bond
(C) a metallic bond
(D) a coordinate covalent bond

97. The following is a Lewis structure for nitrous acid:

$$H - \ddot{O} - \ddot{N} = \ddot{O}:$$

What is the correct sequence (left to right) of the formal charges?

(A) $+1, -2, +3, -2$
(B) $0, 0, 0, 0$
(C) $+1, -1, +2, -2$
(D) $-1, -2, +5, -2$

98. Referring to the Lewis structure for nitrous acid in the previous question, what is the correct hybridization (left to right) of the oxygen atoms?

(A) sp^3 and sp
(B) sp and sp^2
(C) sp^3 and sp^3
(D) sp^3 and sp^2

99. Referring to the Lewis structure for nitrous acid in question 97, how do the lengths of the nitrogen-oxygen bonds compare when nitrous acid loses a hydrogen ion to form a nitrite ion?

(A) They are equal.
(B) The bond to the oxygen on the left is longer.
(C) The bond to the oxygen on the left is shorter.
(D) The relative lengths are always changing.

100. The following is the structure of the pyruvate ion:

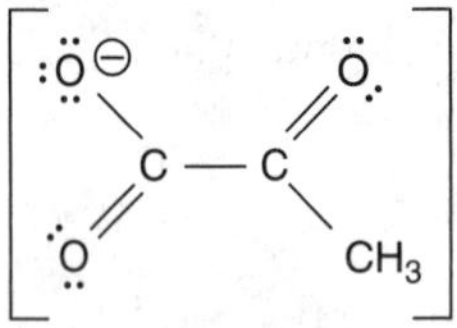

The approximate geometry around the carbon atoms bonded to oxygen (left to right) is

(A) trigonal pyramidal and trigonal planar
(B) trigonal planar and trigonal pyramidal
(C) trigonal planar and trigonal planar
(D) tetrahedral and tetrahedral

101. What is the shape of the phosphate ion?

(A) Trigonal planar
(B) Tetrahedral
(C) Octahedral
(D) Linear

102. Hydrogen chloride gas will react with ammonia gas to produce ammonium chloride. At room temperature, ammonium chloride will exist as what phase?

(A) Gas
(B) Plasma
(C) Liquid
(D) Solid

103. Electrolytes are compounds that produce ions when dissolving in a solvent. Which of the following compounds are NOT electrolytes when dissolved in water?

I. Ethanol
II. Hydrogen bromide
III. Carbon monoxide

(A) I and III only
(B) I and II only
(C) II only
(D) I only

104. Which of the following is a free radical?

(A) NO_2^-
(B) N_2
(C) HCN
(D) NO_2

105. All the $H-C-H$ angles in the methyl radical, CH_3, are 120°. What is the hybridization of the carbon atom?

(A) sp^3
(B) sp
(C) sp^2
(D) Unknown

106. What is the structure of the chlorate ion, ClO_3^-?

(A) Linear
(B) Trigonal planar
(C) Trigonal pyramidal
(D) Bent

107. Glass is primarily silicon dioxide. Hydrofluoric acid is the only common acid that will react with glass. The reaction produces water and silicon tetrafluoride. Which of the following statements about silicon tetrafluoride is true?

(A) It is polar.
(B) It is nonpolar.
(C) It is ionic.
(D) It is explosive.

108. The nitrogen atom in nitric acid (HNO_3) is bonded to all three oxygen atoms. This means that

(A) the nitrogen forms one double bond and two single bonds to the oxygen atoms
(B) the nitrogen forms three single bonds to the oxygen atoms
(C) the nitrogen forms two double bonds and one single bond to the oxygen atoms
(D) the nitrogen atom has a nonbonded pair of electrons

109. Chelation therapy is useful in the treatment of heavy-metal poisoning. The process involves treatment with chelating agents such as BAL (British anti-Lewisite). Chelating agents are ligands that bind to heavy-metal ions through the donation of electron pairs to the metal ion (cation). The bond to the metal is a very polar coordinate covalent bond. Which of the following statements is true?

(A) Calcium ions would be good ligands.
(B) Binding to the chelating agent increases the positive charge on the metal.
(C) Ethane should be a good ligand.
(D) The ligand is a Lewis base.

110. All the halogens except fluorine form acids with the general formula HXO_4 (where X is any halogen other than fluorine). Which of the following does NOT explain why HFO_4 does not exist?

(A) Fluorine is the only halogen that cannot exceed an octet of electrons.
(B) Fluorine has an unusually low ionization energy.
(C) Fluorine cannot form compounds where it has a $+7$ oxidation number.
(D) Fluorine is too small for four oxygen atoms to bond to it.

111. Hydrogen bonds are relatively strong intermolecular forces that are important for many of the properties of water, the double helix of DNA, and the secondary structure of proteins. The hydrogen bonds in water are stronger than the intermolecular interactions of all of the following EXCEPT

(A) the intermolecular forces in NaCl
(B) the intermolecular forces in ammonia
(C) the intermolecular forces in CH_3F
(D) the intermolecular forces in CH_3CH_3

112. The $H-N-H$ bond angle in ammonia is

(A) indeterminate
(B) ideal
(C) greater than ideal
(D) less than ideal

113. What is the shape of the cyanate ion, OCN^-?

(A) Linear
(B) Bent
(C) Trigonal planar
(D) Triangular

114. A burette is a piece of laboratory glassware designed to precisely measure the volume of solution delivered during a titration. Due to capillary action, the solution in the burette forms a meniscus (curved surface) as the solution near the edges climbs up the side of the tube. The cause of the capillary action is the strong intermolecular forces between the glass walls and the solution in the tube. In order to measure the quantity of liquid dispensed, it is necessary to take readings before and after the delivery of a quantity of solution and note the difference. To get accurate readings

(A) both need to be at the top of the meniscus
(B) both need to be at the bottom of the meniscus
(C) both need to be at the average between the top and the bottom of the meniscus
(D) it does not matter if the top or the bottom is used, as long as the choice is consistent

115. What is the shape of a nitrite ion?

(A) Trigonal planar
(B) Linear
(C) Bent
(D) Triangular

116. Electronic transitions in water molecules require photons with energies in the ultraviolet region of the spectrum. Vibrational transitions in water molecules require less energy. Photons causing vibrational transitions have energies in which region of the spectrum?

(A) Ultraviolet
(B) X-ray
(C) Gamma ray
(D) Infrared

117. Infrared spectroscopy examines the vibrational spectra of molecules. It is possible to use infrared spectroscopy to examine solids, liquids, and gases. Which of the following gases has *no* vibrational spectrum?

(A) Xe
(B) H_2O
(C) HCN
(D) NO

118. Carbon dioxide and water are very good greenhouse gases, while oxygen and nitrogen are very poor greenhouse gases. This information indicates that a greenhouse gas needs

(A) to have only single bonds
(B) to be a polar molecule
(C) a permanent dipole moment
(D) polar covalent bonds

119. Freons, such as CHF_2Cl, are useful in air conditioners and in refrigeration systems. The strongest type of intermolecular force in CHF_2Cl is

(A) hydrogen bonding
(B) dipole-dipole
(C) covalent bonding
(D) London dispersion force

120. Phosphorus forms a number of compounds with the halogens. One phosphorus halide is PF_3. Which of the following statements is true concerning PF_3?

(A) The formal charge on phosphorus is 0, and the molecular geometry is trigonal pyramidal.

(B) The formal charge on phosphorus is +3, and the molecular geometry is trigonal pyramidal.

(C) The formal charge on phosphorus is 0, and the molecular geometry is T-shaped.

(D) The formal charge on phosphorus is +3, and the molecular geometry is T-shaped.

121. What is the *expected* trend in the first ionization energies of cadmium, indium, and tin?

(A) Cd > In > Sn

(B) In > Sn > Cd

(C) Sn > In > Cd

(D) In > Cd > Sn

122. Which of the following properties is unlikely to be true for a metal such as copper?

(A) High ductility

(B) High malleability

(C) High electrical conductivity

(D) High brittleness

123. Boron halides are capable of reacting with halide ions via the following general reaction:

$$BX_3 + X^- \rightarrow BX_4^-$$

During this reaction, the boron halide behaves as a

(A) Lewis acid, and the geometry changes from trigonal planar to tetrahedral

(B) Lewis acid, and the geometry changes from trigonal pyramidal to tetrahedral

(C) Lewis base, and the geometry changes from trigonal planar to tetrahedral

(D) Lewis base, and the geometry changes from trigonal pyramidal to tetrahedral

124. Which of the following series indicates a correct order of increasing polarity?

(A) F–F < N–F < C–F
(B) F–F < C–F < N–F
(C) N–F < C–F < F–F
(D) C–F < N–F < F–F

125. The mineral sphalerite, ZnS, has a structure with each zinc atom tetrahedrally surrounded by sulfur, and each sulfur atom tetrahedrally surrounded by zinc. Replacing both the zinc and sulfur with carbon will result in the diamond structure. This means that

(A) London dispersion forces replace ionic bonds
(B) ionic bonds replace covalent bonds
(C) covalent bonds replace ionic bonds
(D) dipole-dipole forces replace covalent bonds

126. Which of the following substances will have the highest heat of vaporization?

(A) H_2O
(B) $NaCl$
(C) Cl_2
(D) HCl

127. Antifreeze (ethylene glycol) is effective at lowering the freezing point of water. The addition of 2 moles of ethylene glycol to 1 L of water will lower the freezing point of the solution to $-3.8°C$. Ideally, what will be the freezing point of a solution made by adding 73 g of HCl to 1 L of water?

(A) $-7.6°C$
(B) $-3.8°C$
(C) $0.0°C$
(D) $-11.4°C$

128. Which of the following has the lowest ionization energy?

(A) Calcium

(B) Lithium

(C) Potassium

(D) Magnesium

129. What general trend is true for any period on the periodic table?

(A) Electronegativity decreases and electron affinity increases with increasing atomic number.

(B) Electronegativity and electron affinity decrease with increasing atomic number.

(C) Electronegativity and electron affinity increase with increasing atomic number.

(D) Electronegativity increases and electron affinity decreases with increasing atomic number.

130. All of the geometries for compounds can be nonpolar EXCEPT

(A) bent

(B) linear

(C) tetrahedral

(D) octahedral

131. Why is it easier for sodium to form cations than anions?

(A) Sodium has a high electron affinity.

(B) Sodium has a high ionization energy

(C) Sodium has a low ionization energy.

(D) Sodium can complete its outer shell by gaining an electron.

132. Potassium metal reacts explosively with all of the following EXCEPT

(A) fluorine gas

(B) argon gas

(C) water

(D) oxygen gas

133. Channels are present in cell membranes to allow the selective transport of ions through the membrane. Size is one of the factors that allow the selective transport of the ions. Membranes are less permeable to larger ions than smaller ions. Which of the following statements is correct?

(A) Sodium ions and chloride ions transport at the same rate because they are the same size.

(B) Smaller sodium ions transport more readily than larger potassium ions.

(C) Calcium ions transport more slowly because they are larger than potassium ions.

(D) Smaller potassium ions transport more readily than larger sodium ions.

134. Which of the following elements has the lowest electron affinity?

(A) Sulfur
(B) Chlorine
(C) Neon
(D) Fluorine

135. The electron configuration of sulfur is $[Ne]3s^23p^4$. What is the electron configuration of a sulfide ion?

(A) $[Ne]3s^23p^2$
(B) $[Ne]3s^23p^5$
(C) $[Ar]$
(D) $[Ne]$

136. When an oxygen molecule binds to the iron in hemoglobin, the iron becomes six-coordinate. The approximate geometry of the atoms around the iron is

(A) trigonal bipyramidal
(B) square planar
(C) tetrahedral
(D) octahedral

137. Which of the following amino acid residues is capable of forming a coordinate covalent bond to an iron ion?

(A) Aspartic acid

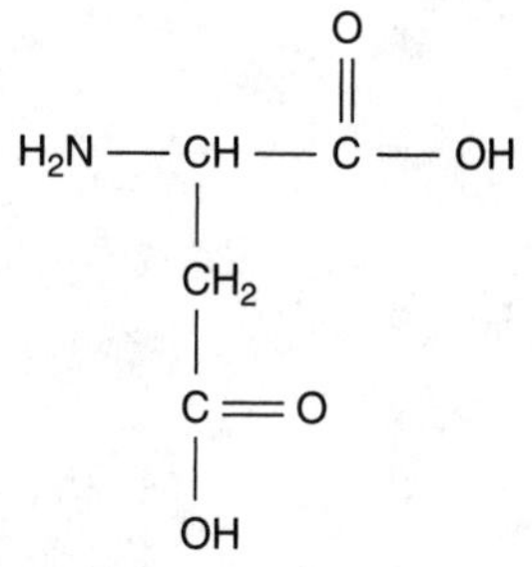

$$H_2N - CH - \overset{\displaystyle O}{\overset{\displaystyle \|}{C}} - OH$$

$$\underset{\displaystyle \begin{array}{c} | \\ CH_2 \\ | \\ C=O \\ | \\ OH \end{array}}{}$$

(B) Valine

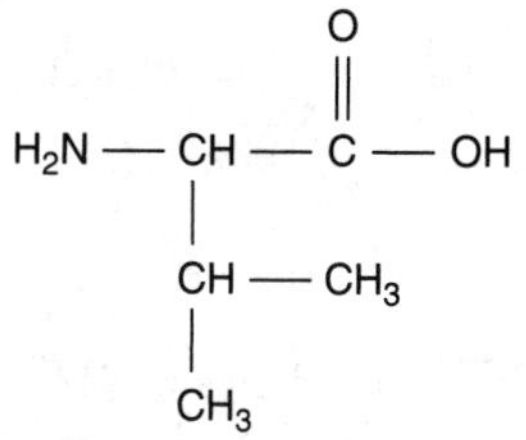

$$H_2N - CH - \overset{\displaystyle O}{\overset{\displaystyle \|}{C}} - OH$$

$$\underset{\displaystyle \begin{array}{c} | \\ CH - CH_3 \\ | \\ CH_3 \end{array}}{}$$

(C) Alanine

$$H_2N - CH - \overset{\displaystyle O}{\overset{\displaystyle \|}{C}} - OH$$

$$\underset{\displaystyle \begin{array}{c} | \\ CH_3 \end{array}}{}$$

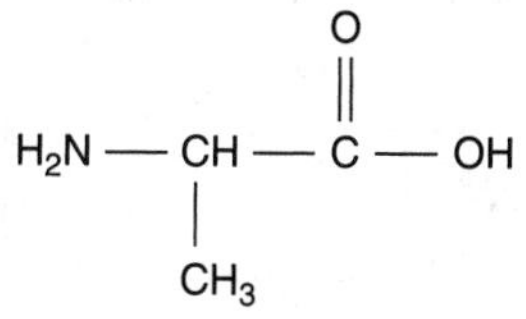

(D) Glycine

$$H_2N - CH - \overset{\displaystyle O}{\overset{\displaystyle \|}{C}} - OH$$

$$\underset{\displaystyle \begin{array}{c} | \\ H \end{array}}{}$$

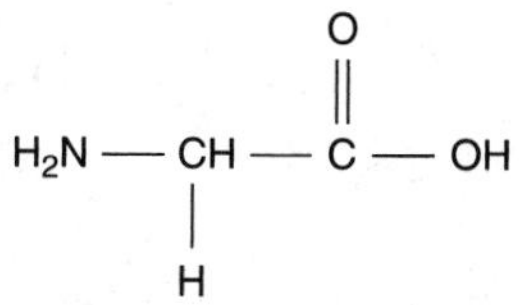

138. In which of the following resonance forms of the cyanate ion does the nitrogen atom have a formal charge of 0?

(A) 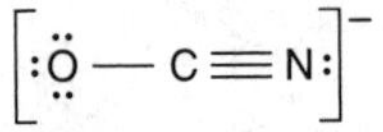 $\left[:\ddot{O} - C \equiv N: \right]^{-}$

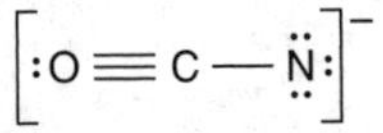

(B) $\left[:O \equiv C - \ddot{N}: \right]^{-}$

(C) $\left[\ddot{O} = C = \ddot{N} \right]^{-}$

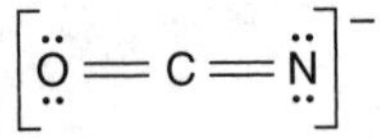

(D) 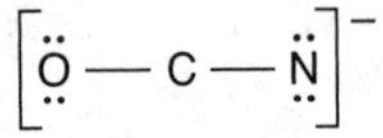 $\left[\ddot{O} - C - \ddot{N} \right]^{-}$

139. The following is the structure of the pyruvate ion:

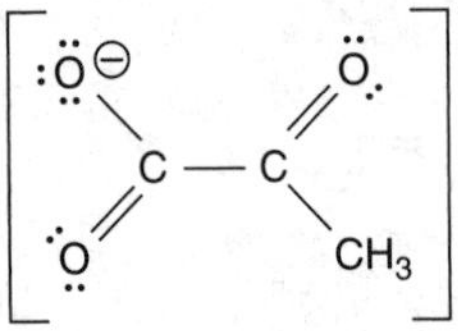

Which of the following statements about the pyruvate ion is true?

(A) Upon lowering the pH, the product will be less soluble.
(B) Upon lowering the pH, the product will be more soluble.
(C) The longest carbon-oxygen bond is to the oxygen atom in the upper right of the structure.
(D) The carbon-oxygen bonds on the left side of the structure are of unequal lengths.

140. Which of the following molecules adopts a trigonal pyramidal geometry?

(A) Sulfur trioxide
(B) Water
(C) Methane
(D) Ammonia

141. Carbon dioxide will dissolve in water to produce carbonic acid. During this process, the geometry around the carbon atom changes from

(A) linear to trigonal planar
(B) bent to trigonal planar
(C) linear to tetrahedral
(D) trigonal planar to trigonal planar

142. Mercury and mercury compounds are toxic. In general, nonpolar dimethyl mercury is more toxic than ionic compounds such as mercury sulfide. Why is dimethyl mercury more toxic?

(A) Nonpolar dimethyl mercury can cross the hydrophobic cell membrane more easily than ionic mercury.

(B) Mercury ions can cross the hydrophobic cell membrane more easily than dimethyl mercury.

(C) Most ionic mercury compounds are soluble and are easily eliminated.

(D) The high pH in the small intestine enhances the absorption of dimethyl mercury.

143. Which of the following compounds will have the lowest melting point?

(A) CH_3CH_2COOH

(B) $CH_3CH_2CH_2OH$

(C) CH_3CH_2CHO

(D) $CH_3CH_2CH_2NH_2$

CHAPTER 4

Phases

144. The vapor pressure of a liquid varies with temperature. What happens when the vapor pressure of the liquid equals the external pressure?

(A) It explodes.
(B) It boils.
(C) It freezes.
(D) It dissociates.

145. The boiling point of water is lower in Denver, Colorado, than in Houston, Texas. Which of the following explains this difference?

(A) An increase in altitude involves an increase in atmospheric pressure.
(B) An increase in altitude involves a decrease in atmospheric pressure.
(C) An increase in altitude involves a decrease in temperature.
(D) An increase in altitude involves an increase in temperature.

146. The heat of vaporization of water is very high. The reason the value is high is that

(A) the hydrogen bonds are strong
(B) the $O-H$ bonds in water are strong
(C) water is a very light molecule
(D) the ionic bonds between H^+ and OH^- are strong

147. A substance is a gas at STP. Decreasing the temperature at constant pressure may eventually cause

(A) evaporation
(B) sublimation
(C) deposition
(D) fusion

148. The normal boiling point of nitrogen is 77 K. However, it is possible to have liquid nitrogen as a stable phase at temperatures above 77 K. Why is this possible?

    (A) The pressure is in equilibrium with the gas.
    (B) The pressure is less than 1 atm.
    (C) The liquid is in equilibrium with the solid.
    (D) The pressure is greater than 1 atm.

149. Ethanol has a normal boiling point of 78°C. Ethanol boils at 78°C under which of the following conditions?

    (A) When the pressure is 1 atmosphere
    (B) At any pressure
    (C) When its molar volume is 22.4 L
    (D) At STP

150. Even though diamond and graphite both consist of carbon, diamond is one of the hardest substances known, while graphite is one of the softest. Diamond is a three-dimensional network of sp^3 hybridized carbon atoms with each carbon atom bonded to four others. Graphite, on the other hand, has a planar network of sp^2 hybridized carbon atoms with each bonded to three others in the plane through a resonating network. An explanation of why graphite is softer than diamond is that

    (A) graphite is resonance stabilized
    (B) sp^2 bonds are weaker than sp^3 bonds
    (C) graphite has only weak London dispersion forces between the planes of carbon atom instead of strong covalent bonds
    (D) the planes in graphite form a less compact structure than diamond

151. The molar heat of vaporization is the amount of energy necessary to vaporize a mole of a substance. Which of the following will have the highest heat of vaporization?

    (A) Ethanol
    (B) Formaldehyde
    (C) Propane
    (D) Hydrogen sulfide

152. Ozone forms in the stratosphere. This ozone helps protect the surface of the earth from harmful ultraviolet radiation. The pressure in the stratosphere where the ozone is forming is about 8 torr. Which of the following statements is true?

    (A) The boiling point of water is lower at 8 torr.
    (B) The partial pressure of ozone in the stratosphere is 8 torr.
    (C) The boiling point of water is a constant 100°C.
    (D) The energy of the collisions of the gas molecules is lower at lower pressures.

153. All of these processes release energy to the surroundings EXCEPT

    (A) freezing
    (B) deposition
    (C) condensation
    (D) sublimation

154. A scientist needs to know the heat capacity of a solid for a calorimeter. If she has only the specific heat, what must she do?

    (A) Divide the specific heat by the molar mass of the solid.
    (B) Divide the specific heat by the mass.
    (C) Multiply the specific heat by the mass.
    (D) Multiply the specific heat by the molar mass of the solid.

155. The following is a generalized heating curve showing the transition of a substance from the solid state to the gaseous state. This diagram shows the temperature of the sample versus the heat added:

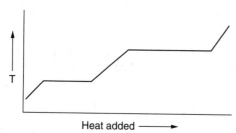

T

Heat added ⟶

    Calculations concerning the "slanting" parts of the graph require information on which of the following?

    (A) Heat of fusion
    (B) Heats of vaporization
    (C) Specific heats
    (D) Heat of sublimation

156. The sum of the heat of fusion plus the heat of vaporization is equivalent to the

    (A) specific heat of the substance
    (B) heat of deposition
    (C) heat of evaporation
    (D) heat of sublimation

157. The following is a generalized heating curve showing the transition of a substance from the solid state to the gaseous state. This diagram shows the temperature of the sample versus the heat added:

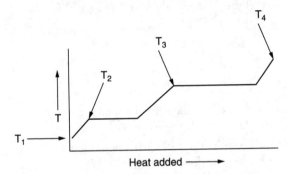

When calculating the energy necessary to heat a sample from the melting point to the boiling point, it is necessary to know the amount of sample, the specific heat of the liquid, and

    (A) T_2 and T_3
    (B) T_1 and T_3
    (C) T_1 and T_4
    (D) T_2 and T_4

158. The heat of fusion for ice is 6.0 kJ/mol, and the molar heat capacity of water is 75 J/mol°C. What is the temperature change when 6.0 kJ of heat is added to 36 g of ice?

    (A) 80°C
    (B) 0°C
    (C) 40°C
    (D) 60°C

159. The phase diagram for water is shown here:

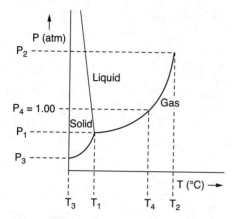

The pressure and temperature at the critical point are

(A) P_3 and T_3
(B) P_1 and T_1
(C) P_2 and T_2
(D) P_4 and T_4

160. How do the heats of fusion for the following substances compare?

NaCl CCl_4 H_2O Al_2O_3

(A) $Al_2O_3 > NaCl > H_2O > CCl_4$
(B) $NaCl > CCl_4 > H_2O > Al_2O_3$
(C) $H_2O > CCl_4 > Al_2O_3 > NaCl$
(D) $CCl_4 > H_2O > NaCl > Al_2O_3$

161. A sodium chloride solution is boiled. The predominant species in the vapor phase is/are

(A) $H_2O(g)$ and $NaCl(g)$
(B) $H_2(g)$ and $O_2(g)$
(C) $H_2O(g)$
(D) $H_2(g)$, $O_2(g)$, and $NaCl(g)$

162. Below is the phase diagram for water:

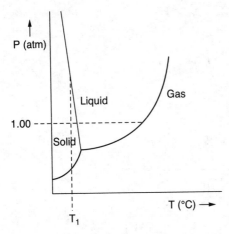

A sample of water vapor at T_1 is compressed at constant temperature. The sequence of phase changes is

(A) solid → liquid → gas
(B) gas → liquid → solid
(C) gas → solid → liquid
(D) No phase change occurs.

163. Even though diamond and graphite both consist of carbon, diamond is the hardest substance known, while graphite is one of the softest. However, the heat of sublimation of both substances is very high. Why is this so?

(A) In both cases, it is necessary to break strong covalent bonds.
(B) In both cases, it is necessary to break strong ionic bonds.
(C) In both cases, it is necessary to break the unusually strong London dispersion forces.
(D) In both cases, it is necessary to break strong metallic bonds.

164. Consider the following thermochemical equations:

$$2 H_2(g) + O_2(g) \rightarrow 2 H_2O(l) \qquad\qquad \Delta H = -570 \text{ kJ}$$
$$N_2O_5(g) + H_2O(l) \rightarrow 2 HNO_3(l) \qquad\qquad \Delta H = -80 \text{ kJ}$$
$$N_2(g) + 3 O_2(g) + H_2(g) \rightarrow 2 HNO_3(l) \qquad\qquad \Delta H = -350 \text{ kJ}$$

What is the value of ΔH for the following reaction?

$$2 N_2(g) + 5 O_2(g) \rightarrow 2 N_2O_5(g)$$

(A) -30 kJ
(B) $-3{,}390$ kJ
(C) 30 kJ
(D) $3{,}390$ kJ

165. Determine the specific heat of a metal sample if a 40 g sample absorbs 240 J of heat while the temperature rises from 25.0°C to 85.0°C.

(A) 0.02 J/g K
(B) 0.10 J/g K
(C) 0.01 J/g K
(D) 0.20 J/g K

166. A 0.46 g sample of ethanol, C_2H_5OH, was placed in a bomb calorimeter and ignited according to the following equation:

$$C_2H_5OH(l) + 3 O_2(g) \rightarrow 2 CO_2(g) + 3 H_2O(l)$$

The total heat capacity of the calorimeter was 1.2 kJ/°C, and the reaction increased the temperature of the system from 22.50°C to 32.50°C. Determine the heat of reaction in kilojoules per mole of ethanol.

(A) -1.2×10^3 kJ
(B) $+1.2 \times 10^3$ kJ
(C) $+2.4 \times 10^3$ kJ
(D) -3.0×10^5 kJ

167. When an 8.6 g sample of sodium nitrate, $NaNO_3$, dissolved in 150.0 g of water in a coffee-cup calorimeter, the temperature changed from 25.3°C to 22.3°C. Calculate the enthalpy change, in kJ/mol $NaNO_3$, for sodium nitrate dissolving in water. Assume that the specific heat of the solution is the same as that of pure water (4.2 J/g K).

(A) -20 kJ/mol
(B) $+20$ kJ/mol
(C) $+1.8 \times 10^3$ kJ/mol
(D) -1.8×10^3 kJ/mol

168. Hydrogen iodide, HI, melts at $-50°C$ and boils at $-35°C$. For this compound, the enthalpy of fusion is 2.9 kJ/mol, and the enthalpy of vaporization is 44 kJ/mol. The molar specific heats for the liquid and gaseous phases are 44 J/mol·K and 25 J/mol·K, respectively. How much energy is necessary to convert 260 g of solid HI at the melting point to the vapor phase at 0.0°C?

(A) $+100$ kJ
(B) $+140$ kJ
(C) $+3,300$ kJ
(D) $+1,000$ kJ

169. Potassium metal reacts with water according to the following equation:

$$2 \text{ K}(s) + 2 \text{ H}_2\text{O}(l) \rightarrow 2 \text{ KOH}(aq) + \text{H}_2(g) \qquad \Delta H° = -130 \text{ kJ}$$

If the standard heat of formation for $H_2O(l)$ is -286 kJ/mol, what is the standard heat of formation for KOH(aq)?

(A) $+350$ kJ/mol
(B) -700 kJ/mol
(C) -350 kJ/mol
(D) Impossible to determine

170. Water boils at a lower temperature at the top of a mountain than at sea level. The BEST explanation for this fact is that

(A) ultraviolet light breaks the H–O bonds
(B) there is more solar energy available at higher altitudes
(C) the vapor pressure of water decreases with increasing altitude
(D) the atmospheric pressure is lower

171. Which of the following is the chief factor leading to the density of ice being less than the density of water?

(A) Hydrogen bonding
(B) Formation of covalent bonds between the water molecules
(C) The energy change when water freezes
(D) The lower entropy of the solid

172. Why does water have a higher molar heat of vaporization than methyl chloride?

(A) The molecular weight of methyl chloride is higher than that of water.
(B) The H−O bonds in water are stronger than any of the bonds in methyl chloride.
(C) Dipole-dipole forces in methyl chloride are weaker than the intermolecular forces in water.
(D) Chlorine causes methyl chloride to be more polar than water.

173. The heat of vaporization of acetone is 0.500 kJ/g, and the boiling point of acetone is 56°C. How much heat is necessary to vaporize 100.0 g of acetone at 329 K?

(A) 1.36×10^4 kJ
(B) 50.0 kJ
(C) 100.0 kJ
(D) Not possible to determine

174. During an experiment, a graduate student adds 100 J of energy to a sample of ethanol at a temperature below the boiling point. The addition of energy causes no change in temperature. Which of the following could explain this observation?

(A) More energy is necessary to change the temperature.
(B) The sample is at the melting point.
(C) The bond energies are all greater than 100 J, so no bonds could be broken.
(D) The energy dissipates as friction.

175. The boiling point of ethanol is 78°C. At the boiling point, there is a phase change from liquid to gas. The enthalpy change for this process is

(A) positive, because *intra*molecular forces are being overcome
(B) positive, because *inter*molecular forces are being overcome
(C) negative, because *inter*molecular forces are being overcome
(D) negative, because *intra*molecular forces are being overcome

176. When water freezes, there is a nearly 10% increase in volume. This increase in volume may result in the bursting of the cooling system in an automobile or the cracking of a sidewalk during winter. Which of the following substances would be the LEAST effective at preventing the cracking of sidewalks?

    (A) $C_6H_{12}O_6$
    (B) $NaCl$
    (C) $(NH_4)_2SO_4$
    (D) Na_3PO_4

177. The following is a simplified phase diagram of water:

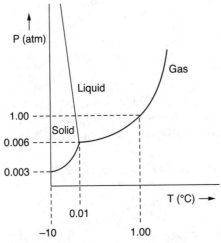

What process occurs when heating a sample of water from $-10°C$ to $+10°C$ at a pressure of 4×10^{-3} atm?

    (A) Deposition
    (B) Sublimation
    (C) Melting
    (D) Boiling

178. The following standard enthalpies of formation are reported for sulfur trioxide (SO_3):

$SO_3(s)$ -454 kJ/mol
$SO_3(l)$ -441 kJ/mol
$SO_3(g)$ -396 kJ/mol

What is the standard heat of deposition for sulfur trioxide?
(A) $+58$ kJ/mol
(B) -58 kJ/mol
(C) -13 kJ/mol
(D) $+13$ kJ/mol

179. The following is a generalized heating curve showing the transition of a substance from the solid state to the gaseous state. This diagram shows the temperature of the sample versus the heat added:

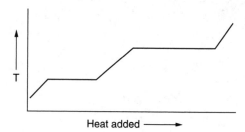

Heat added ⟶

In most cases, the second horizontal region is longer than the first (at a constant heating rate). Why is the second horizontal region longer?
(A) The heat of vaporization is greater than the heat of fusion.
(B) The heat of fusion is greater than the heat of vaporization.
(C) The heat of vaporization is greater than the heat of sublimation.
(D) The heat of fusion is greater than the heat of condensation.

180. The specific heat of water is 4.2 J/g°C. What is the heat capacity of 20 g of water?
(A) 84 J/g°C
(B) 84 J/°C
(C) 42 J/g°C
(D) 42 J/°C

181. A 25 g sample of a substance was dissolved in 175 g of water. The temperature of the resultant solution was 10.0°C higher than the original water. Assuming the specific heat of the solution was the same as pure water (4.2 J/g°C), what was the enthalpy change for the process?
    (A) −8.4 kJ
    (B) +8.4 kJ
    (C) −1.1 kJ
    (D) +7.4 kJ

182. Above the critical point on a phase diagram, the
    (A) gas ceases to exist
    (B) liquid ceases to exist
    (C) liquid vaporizes
    (D) liquid and gas phases are indistinguishable

CHAPTER 5

Gases

183. A mixture of gases has a total pressure of 1,000 torr. The mixture contains 25% methane, 40% oxygen, and 35% hydrogen. What is the partial pressure of hydrogen?

(A) 400 torr
(B) 35 torr
(C) Impossible to determine
(D) 350 torr

184. A steel cylinder contains 1 mole of an ideal gas. At constant temperature, increasing the moles of gas will cause

(A) the pressure to decrease
(B) the pressure to increase
(C) no change in the pressure
(D) an undetermined change in the pressure unless the actual volume is known

185. Air is primarily nitrogen and oxygen with small amounts of other gases. Which of the following statements is true concerning the nitrogen and oxygen in air at room temperature?

(A) Oxygen is heavier, so its molecules must move faster to compensate.
(B) Oxygen is heavier, so it has a higher kinetic energy.
(C) Both gases have the same kinetic energy.
(D) Kinetic energy determinations are not possible, due to interference by the other gases.

186. Many shock absorbers rely on air to function. Which of the following reasons is the BEST reason why water could not replace the air in a shock absorber?

(A) Water freezes as the pressure increases.
(B) Water freezes at 32°F.
(C) Water is denser than air.
(D) Water is nearly incompressible.

187. How will the absolute temperature change when doubling the pressure on a sample of an ideal gas at constant volume?

(A) It remains constant.
(B) It halves.
(C) It doubles.
(D) The answer depends on the number of moles.

188. A number of equations are applicable to the behavior of gases. The basis used for the explanation of the diffusion of gases is

(A) $\left(P + \dfrac{an^2}{v^2}\right)(V - nb) = nRT$

(B) $P_1V_1 = P_2V_2$

(C) $PV = nRT$

(D) $KE = \frac{1}{2}\, mv^2$

189. The following apparatus was used to examine gas phase reactions. The two compounds react to form a nonvolatile compound, which is identified by the formation of a cloud.

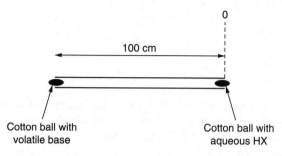

In one experiment, the volatile base was dimethylamine, $(CH_3)_2NH$, and the cloud formed very near the center. The identity of the other reactant might be

(A) HI
(B) HF
(C) HCOOH
(D) HSCN

190. Which of the following gases is most likely to be ideal?

(A) Xe
(B) He
(C) HF
(D) C_2H_6

191. A student builds the apparatus shown here:

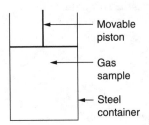

In an experiment, the student began with a 0.50 mole sample of argon gas at 0°C and 1.00 atm. The initial volume was 11 L. The sample was heated at constant pressure to 273°C, and the final volume was determined. What was the final volume?

(A) 11 L
(B) 273 × 11 L
(C) 22 L
(D) 44 L

192. A student builds the apparatus shown in question 191. In an experiment, the student began with a 0.50 mole sample of argon gas at 0°C, 1.00 atm, and the piston was locked into place. The sample was then heated to 273°C and the final volume determined. How do the final pressure (P_f) and volume (V_f) relate to the initial pressure (P_i) and volume (V_i)?

(A) $P_f = 2\ P_i$ and $V_f = 2\ V_i$
(B) $P_f = P_i$ and $V_f = V_i$
(C) $P_f = 2\ P_i$ and $V_f = V_i$
(D) $P_f = P_i$ and $V_f = 2\ V_i$

193. A student builds the apparatus shown in question 191. In an experiment, the student began with a 0.50 mole sample of argon gas at 0°C, 1.00 atm, and the piston was locked into place. The sample was then heated to 273°C, and the initial pressure (P_i) and volume (V_i) were measured. At this point, the piston is unlocked, and a heater/cooler keeps the temperature constant. How do the final pressure (P_f) and volume (V_f) relate to the initial pressure (P_i) and volume (V_i)?

(A) $P_f < P_i$ and $V_f > V_i$
(B) $P_f < P_i$ and $V_f < V_i$
(C) $P_f > P_i$ and $V_f > V_i$
(D) $P_f = P_i$ and $V_f > V_i$

194. A sample of a gas mixture contains helium, neon, and argon. The partial pressures of these gases are He = 0.50 atm, Ne = 76 torr, and Ar = 10 kPa. What is the total pressure of the mixture?

(A) 0.70 atm
(B) 10 kPa
(C) 0.50 atm
(D) 1.0 atm

195. A mixture of gases made of 16.0 g of CH_4, 30.0 g of C_2H_6, and 58.0 g of C_4H_{10} was contained in a flask. The total pressure of the gases was 1.20 atm. Determine the partial pressure of each gas, in atmospheres.

(A) CH_4 = 0.20 atm, C_2H_6 = 0.35 atm, and C_4H_{10} = 0.65 atm
(B) CH_4 = 0.40 atm, C_2H_6 = 0.40 atm, and C_4H_{10} = 0.40 atm
(C) CH_4 = 0.15 atm, C_2H_6 = 0.25 atm, and C_4H_{10} = 0.80 atm
(D) Impossible to determine without knowing the volume of the flask

196. A sample of a gas contains 1.0 mole of a gas at 1.0 atmosphere and 0°C. The density of the gas is 2.0 g/L. What might be the identity of the gas?

(A) CO_2
(B) CH_4
(C) H_2
(D) Xe

197. A sample of a gas is held at constant pressure. Initially, the gas occupied 500 mL at a temperature of 25°C. The temperature was changed until the volume became 250 mL. What was the temperature at the new volume?

(A) 13°C
(B) −120°C
(C) 120°C
(D) −13°C

198. A student has two balloons. Both balloons have the same volume, pressure, and temperature. One of the balloons contains hydrogen gas, and the other balloon contains oxygen gas. How do the numbers of molecules in each balloon compare?

(A) There are more oxygen molecules.
(B) There are more hydrogen molecules.
(C) They are the same.
(D) There is no way to determine the number of molecules.

199. A student has two balloons. Both balloons have the same volume, pressure, and temperature. One of the balloons contains hydrogen gas, and the other balloon contains oxygen gas. How do the speeds and average kinetic energies of molecules in each balloon compare?

(A) The hydrogen molecules are moving faster, but they have the same average kinetic energy as the oxygen molecules.

(B) The hydrogen molecules are moving slower, but they have the same average kinetic energy as the oxygen molecules.

(C) The hydrogen molecules are moving faster, and they have a lower average kinetic energy than the oxygen molecules.

(D) The hydrogen molecules are moving faster, and they have a higher average kinetic energy than the oxygen molecules.

200. The rate of effusion for an unknown gas was found to be half that of methane gas under the same conditions. Which of the following might be the unknown gas?

(A) O_2
(B) He
(C) CO_2
(D) SO_2

201. The van der Waals constants for three gases are given in the following table:

Gas	a (L² atm/mol²)	b (L/mol)
CO_2	3.658	0.04286
SO_2	6.865	0.05679
O_2	1.382	0.03186

Which of the gases is the closest to an ideal gas?

(A) SO_2
(B) CO_2
(C) O_2
(D) All are equally close.

202. Gas samples approach ideal behavior at low pressure and high temperature. Which postulate of kinetic molecular theory is a problem at high pressure?

(A) A gas is composed of particles called atoms or molecules.
(B) The volumes of the gas particles are negligible.
(C) The average kinetic energy of the molecules depends on the absolute temperature.
(D) The particles are constantly moving.

203. Gas samples approach ideal behavior at low pressure and high temperature. Which postulate of kinetic molecular theory is a problem at low temperature?

(A) There are no attractive or repulsive interactions between the gas particles.
(B) The volumes of the gas particles are negligible.
(C) The average kinetic energy of the molecules depends on the absolute temperature.
(D) The particles are constantly moving.

204. Two gas samples initially have the same volume (500 mL), temperature (27°C), and pressure (1 atm). The pressure on one sample is halved, while the absolute temperature of the other sample is doubled. What are the relative volumes of the two samples after the changes?

(A) The lower-pressure sample is larger.
(B) The heated sample is larger.
(C) They are both the same.
(D) The lower-pressure sample is smaller.

205. A student does an experiment examining the properties of water vapor. During each step of the experiment, she finds that the pressure of the water vapor is a little lower than expected from $PV = nRT$. Why is the pressure low?

(A) There is a strong attraction between the water molecules.
(B) The water molecules are not linear in the gas phase.
(C) The gaseous water molecules are nearly ideal.
(D) There was some liquid water present in the sample.

206. The volume of 2 moles of N_2 at STP is

(A) 56.0 L
(B) 28.0 L
(C) 44.8 L
(D) 22.4 L

207. A gaseous mixture containing helium, neon, and argon has a total pressure of 100 torr. The partial pressure of helium is 35 torr, and the partial pressure of neon is 15 torr. What percentage of the total pressure is due to argon?

(A) 33%
(B) Impossible to determine
(C) 50%
(D) 25%

208. A sample contains two gases, A and B. The velocity of gas B is twice that of gas A. What are the possible identities of gases A and B?

(A) Kr and He
(B) O_2 and N_2
(C) H_2 and Xe
(D) Kr and Ne

209. A latex balloon contains a mixture of helium contaminated with air. The helium causes the balloon to rise. The next day, the balloon does not rise to the same degree as previously. Why?

(A) The light helium rapidly effuses out of the balloon.
(B) Heavier air molecules effuse into the balloon.
(C) The helium molecules slow down overnight, which lowers the temperature and causes the balloon to contract.
(D) Water vapor effuses into the balloon and condenses.

210. A sample of a gas is confined in a cylinder with a moving piston. If the diameter of the cylinder is halved and the force on the gas doubles, how will the volume change?

(A) No change
(B) Increase
(C) Decrease
(D) Unknowable without knowing the temperature

211. The following apparatus was used to examine gas phase reactions. The two compounds react to form a nonvolatile compound, which is identified by the formation of a cloud.

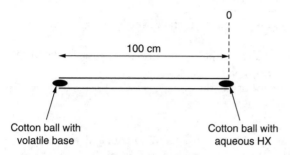

In one experiment, the volatile base was methylamine, CH_3NH_2, and a series of HX materials was used, which yielded the following data:

Sample	Position of cloud (relative to 0)
HX1	10 cm
HX2	17 cm
HX3	23 cm

According to these results, HX1

(A) has the highest molecular weight
(B) has the highest vapor pressure
(C) is the strongest acid
(D) has the lowest activation energy

212. A student builds the apparatus shown here:

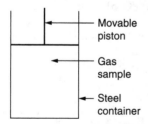

In an experiment, the student began with a 0.50 mole sample of argon gas at 0°C and 1.00 atm. The sample was heated at constant pressure to 273°C and the final volume determined. What was the initial volume?

(A) $1.1 \times 10^{-2} \text{ m}^3$
(B) $2.2 \times 10^{-2} \text{ m}^3$
(C) $5.6 \times 10^{-3} \text{ m}^3$
(D) $4.4 \times 10^{-2} \text{ m}^3$

213. Which of the following is NOT based on an assumption of kinetic molecular theory?

   (A) The intermolecular forces between the molecules of all gases are negligible.
   (B) At a given temperature, the average kinetic energy of all gas molecules is the same.
   (C) The volume of all gas molecules is negligible compared with the volume of the container.
   (D) At a given temperature, the average velocity of all gas molecules is the same.

214. A sample of a gas mixture contains helium, neon, and argon. The partial pressures of these gases are He = 0.50 atm, Ne = 0.20 atm, and Ar = 0.30 atm. What are the mole fractions of each gas?

   (A) He = 50, Ne = 20, Ar = 30
   (B) He = 1.0, Ne = 0.40, Ar = 0.60
   (C) He = 0.50, Ne = 0.20, Ar = 0.30
   (D) Impossible to determine without knowing either the volume or the temperature

215. A sample of a gas contains 1.0 mole of gas at 1.0 atmosphere pressure and 77°C. What is the volume of the gas?

   (A) 6.2 L
   (B) 22.4 L
   (C) 28.7 L
   (D) 20.1 L

216. A sample of a volatile liquid is placed in an 80 mL container and volatilized. The vapor exerts a pressure of 3.0 atm at a temperature of 27°C. The mass of the vapor was 0.6 g. What might be the identity of the volatile liquid?

   (A) CCl_4
   (B) C_8H_{18}
   (C) C_4H_{10}
   (D) CH_3OH

217. A student used the following reaction to generate hydrogen gas:

$$Zn(s) + H_2SO_4(aq) \rightarrow ZnSO_4(aq) + H_2(g)$$

The student collected the gas by water displacement. The complete reaction of 0.010 mole of zinc with excess sulfuric acid generated 224 mL of gas at 0°C. The student used the ideal gas equation to calculate the pressure of the gas. The calculated pressure was 1.0 atm. However, when the student measured the pressure, he found that it was greater than calculated. Which of the following BEST explains the discrepancy?

(A) The student did not include the vapor pressure of water.
(B) The student forgot to account for the volatility of sulfuric acid.
(C) The kinetic energy of the hydrogen molecules is higher at the freezing point of water.
(D) The student did not consider the formation of ice in the system.

CHAPTER 6

Solutions

218. A sample of hydrogen gas was collected over water. Only part of the pressure of the collected gas was due to hydrogen. The remaining pressure was due to

(A)   water vapor
(B)   oxygen gas
(C)   liquid hydrogen
(D)   inert gases

219. Why does $NaNO_2$ produce a greater freezing point depression than HNO_2?

(A)   Only $NaNO_2$ dissociates completely in solution.
(B)   Only HNO_2 dissociates completely in solution.
(C)   The heat of solution for $NaNO_2$ is more endothermic than that of HNO_2.
(D)   The heat of solution for HNO_2 is more endothermic than that of $NaNO_2$.

220. Water has a vapor pressure of 32 torr at 30°C. What will be the vapor pressure if 10.0 g of magnesium chloride is added to 100.0 mL of water at 30°C?

(A)   It will be lower.
(B)   It will be higher.
(C)   It will be the same.
(D)   It will be the average of the vapor pressure of the two substances.

221. Sulfur dioxide gas (SO_2) is most soluble in which of the following solutions?

(A) 1.0 M NaOH
(B) 1.0 M Na_2SO_3
(C) 2.0 M Na_2SO_3
(D) 1.0 M HCl

222. Two grams of an unknown salt was added to 100.0 mL of pure water. The salt completely dissolved. Which of the following might be the salt?

(A) Ag_2S
(B) $CaCO_3$
(C) $AlPO_4$
(D) $Ca(C_2H_3O_2)_2$

223. The vapor pressure of water does NOT depend on

(A) the external pressure
(B) the temperature
(C) hydrogen bonding
(D) the presence of solutes

224. Hydrogen chloride gas is very soluble in water. Which of the following statements is true concerning the solution formed?

(A) The hydrogen chloride exists primarily as molecules.
(B) The boiling point is higher than that of pure water.
(C) The greater the concentration of hydrogen chloride, the higher the pH.
(D) None of the above

225. Why is decanoic acid ($C_9H_{19}COOH$) more soluble in a basic solution than in pure water?

(A) The acid reacts with the base to produce a hydrophilic ion.
(B) The acid reacts with the base to produce a hydrophobic ion.
(C) Weak ion-dipole forces replace the hydrogen bonds present in the pure acid.
(D) A basic solution contains hydrated ions that are not present in pure water.

226. Which of the following solutions will boil at the highest temperature?

(A) 1.0 M HNO$_3$
(B) 1.0 M NaCl
(C) 2.0 M CH$_3$CH$_2$OH
(D) 1.0 M CaCl$_2$

227. The freezing point of a solution is depressed by an amount determined by the relationship $\Delta T = iK_f m$, where i is the van't Hoff factor (the number of mole particles produced by each mole of solute), K_f is the solvent-specific freezing point depression constant, and m is the molality of the solution. A solution is found to have $\Delta T = -0.80°C$. Before the researcher can repeat the experiment, he spills half of the solution. However, an attempt is made to determine the freezing point of the remaining solution. The freezing point of the remaining solution is

(A) $-0.40°C$
(B) $-0.80°C$
(C) $-1.60°C$
(D) impossible to determine

228. Salts such as sodium chloride and calcium chloride lower the freezing points of aqueous solutions. The reason why this occurs is that

(A) it is necessary to cool the solution further to compensate for the heat of solution of the salts
(B) the dissolved salts increase the density of the solution
(C) the ions from the dissolved salts interfere with the formation of ice crystals
(D) the dissolved salts lower the vapor pressure of the solution

229. In osmosis, a semipermeable membrane separates two solutions. Which of the following statements is true concerning osmosis?

(A) The osmotic pressures of a 1.0 m sodium chloride solution is the same as that of a 1.0 m acetic acid solution.
(B) The solvent does not flow if the concentrations are the same in both solutions.
(C) The solvent always flows into the solution of higher concentration.
(D) Osmotic pressure forces the solvent through the membrane.

230. The van't Hoff factor, i, is a term used to express the number of particles produced by an electrolyte in solution. The value of this factor may vary due to interactions between the particles. Under which of the following conditions would the van't Hoff factor be LEAST likely to vary from the ideal value?

(A) Low concentrations
(B) High concentrations
(C) High temperatures
(D) Low pressures

231. A scientist measures the freezing point depression of a 0.1 m sodium nitrate solution. He uses the ideal van't Hoff factor for the calculation of the ideal freezing point depression. The observed freezing point is

(A) not measurable, because the solution did not freeze
(B) lower than the theoretical
(C) less than the change in the boiling point
(D) higher than the theoretical

232. The van't Hoff factor, i, is a term used to express the number of particles produced by an electrolyte in solution. The value of this factor may vary due to interactions between the particles. There are a number of methods available to determine the variation in the van't Hoff factor. Which of the following experiments would NOT lead to a determination of this variation?

(A) A density measurement of the solution
(B) A freezing point measurement of the solution
(C) A boiling point measurement of the solution
(D) An osmotic pressure measurement of the solution

233. Ion-exchange resins, such as water softeners, are capable of removing ions from solution. During the removal, ions in the solution displace ions already on the resin. The exchange occurs because the ions in the solution have a greater affinity for the resin than the ions already present. Most water softeners begin with sodium ions on the resin. A sodium ion water softener would be LEAST effective in removing which of the following ions from solution?

(A) Li^+
(B) K^+
(C) Mg^{2+}
(D) Ca^{2+}

234. Reverse osmosis is capable of removing ions from solution through the application of a pressure greater than the osmotic pressure. Prior to the application of pressure, a reducing agent is often used to treat the water. Which of the following ions will the reducing agent be most likely to affect?

(A) Hg^{2+}
(B) Na^+
(C) Ca^{2+}
(D) Al^{3+}

235. It is possible to purify water by distillation. However, purification by distillation is less effective if the contaminant is volatile. Distillation would work BEST for which of the following contaminants?

(A) $C_2H_2Br_2$
(B) CH_2Cl_2
(C) NCl_3
(D) $HgCl_2$

236. The vapor pressure of water is about 0.030 atm at 25°C. What is the vapor pressure of a solution made by adding 5.0 moles of antifreeze (ethylene glycol) to 1.0 L of water?

(A) 0.040 atm
(B) 0.028 atm
(C) 0.015 atm
(D) 0.0080 atm

237. Four 1 M test solutions (A, B, C, and D) were cooled at equal rates. The identities of the test solutions are A = NaCl, B = $NaNO_3$, C = Na_3PO_4, and D = Na_2S. Which would be the last solution to freeze?

(A) D
(B) C
(C) B
(D) A

238. A precipitate will form when mixing a silver nitrate solution with each of the following solutions: NaCl, NaI, Na_2S, and Na_3PO_4. Which solution will give a precipitate with the strongest ionic bonding?

(A) Na_3PO_4
(B) Na_2S
(C) NaI
(D) NaCl

239. Mixing 50.00 mL of 4.00 M NaOH with 50.00 mL of 4.00 M HNO$_3$ in a calorimeter resulted in a 52°C temperature increase. The freezing point of the final solution would be

(A) higher than that of either of the original solutions
(B) lower than that of either of the original solutions
(C) equal to that of either of the original solutions
(D) the average of the freezing points of the original solutions

240. The experimental measurement of the osmotic pressure of a solution containing Pt(NH$_3$)$_4$Cl$_2$ indicated that the compound ionized into three ions in aqueous solution. What is the most likely geometry of the cationic complex?

(A) Linear
(B) Tetrahedral
(C) Octahedral
(D) Square planar

241. Which of the following is more soluble in 1.0 M HCl?

(A) CaCO$_3$(s)
(B) NaCl(s)
(C) Ca(NO$_3$)$_2$(s)
(D) KBr(s)

242. Which of the following is the most soluble in water?

(A) CH$_3$CH$_2$OH
(B) CH$_3$CH$_2$CH$_2$CH$_2$CH$_2$OH
(C) CH$_3$CH$_2$CH$_2$CH$_2$CH$_2$CH$_2$CH$_2$OH
(D) CH$_3$CH$_2$CH$_2$CH$_2$CH$_2$CH$_2$CH$_2$CH$_2$CH$_2$CH$_2$CH$_2$OH

243. Which of the following properties does NOT depend on the number of particles present in solution?

(A) Vapor pressure
(B) Density
(C) Freezing point
(D) Boiling point

244. If the boiling point constant for water is 0.5°C/m, what is the approximate boiling point of a 2 m solution of magnesium chloride?

(A) 101.5°C

(B) 98.5°C

(C) 97.0°C

(D) 103.0°C

245. Two solutions were prepared. One solution was 1.0 M $NaNO_3(aq)$, and the other was 1.0 M $Ca(NO_3)_2(aq)$. Which solution will freeze at the lower temperature?

(A) Both freeze at the same temperature, because the molarities are equal.

(B) $Ca(NO_3)_2$ will, because there are more ions present in the solution.

(C) $NaNO_3$ will, because there are more ions present in the solution.

(D) $Ca(NO_3)_2$ will, because its molar mass is greater.

246. $NH_3(g)$ and $HCl(g)$ are more soluble in water than $O_2(g)$. Why?

(A) Elements (O_2) are less soluble than compounds (HCl and NH_3).

(B) O_2 has a double bond, which is less reactive than the single bonds in NH_3 and HCl.

(C) NH_3 and HCl are polar and O_2 is not.

(D) O_2 causes combustion, but NH_3 and HCl do not support combustion.

247. Which of the following solutions will behave closest to an ideal solution?

(A) 0.01 M $CaCl_2$

(B) 0.01 M $NaCl$

(C) 0.01 M Na_2CO_3

(D) 0.01 M $Al_2(SO_4)_3$

248. At a given temperature, the vapor pressure of liquid A is greater than that of liquid B. All of the following are true about liquid A EXCEPT which?

(A) Liquid A has a higher boiling point.

(B) Liquid A has a lower boiling point.

(C) Liquid A has weaker intermolecular forces than liquid B.

(D) The freezing point of liquid A is lower than that of liquid B.

249. An important factor leading to oxygen/carbon dioxide exchange in the lungs is

(A) a concentration gradient
(B) a temperature difference
(C) the volume change as the lungs expand
(D) the lower pH of blood in the lungs

250. Which of the following solutions will have the highest boiling point?

(A) 1.0 m $NaNO_3$
(B) 1.0 m $CaCl_2$
(C) 1.0 m $Al(NO_3)_3$
(D) 2.0 m C_2H_5OH

251. The van't Hoff factor, i, is a term used to express the number of particles produced by an electrolyte in solution. The value of this factor may vary due to interactions between the particles. The interactions of the particles in a $CuSO_4$ solution are greater than the interactions of the particles in a KCl solution. Why is this so?

(A) Transition metal ions can form complexes.
(B) Sulfate ions are larger than chloride ions.
(C) The charges of the ions are greater.
(D) The sulfate ion undergoes significant hydrolysis.

252. The van't Hoff factor, i, is a term used to express the number of particles produced by an electrolyte in solution. What is the van't Hoff factor of $(NH_4)_2SO_4$?

(A) 15
(B) 3
(C) 1
(D) 2

253. The first compound isolated during the fractional distillation of the following mixture—PF_3, AsF_3, and SbF_3—would be

(A) AsF_3
(B) PF_3
(C) SbF_3
(D) Fractional distillation will not work on this mixture.

254. Four test solutions (A, B, C, and D) are used to test an unknown. In each case, the concentration of the test solution and the unknown solution was 1 M. The identities of the test solutions are A = NaCl, B = NaNO$_3$, C = NaI, and D = Na$_2$S. The results of mixing each of the test solutions with an unknown were as follows: A (white precipitate), B (no reaction), C (yellow precipitate), and D (black precipitate). The unknown might be

(A) AgNO$_3$
(B) KBr
(C) Ca(NO$_2$)$_2$
(D) Al$_2$(SO$_4$)$_3$

255. An aqueous sodium carbonate solution will react with many aqueous transition metal ion solutions to form a precipitate. Which of the following solutions would be the most effective at redissolving the precipitate?

(A) 1 M C$_2$H$_5$OH
(B) 1 M NaOH
(C) 1 M HNO$_3$
(D) 1 M Na$_2$CO$_3$

CHAPTER 7

Kinetics

256. What happens to the rate of a reaction when the temperature decreases?

(A) The reaction continues at the same rate, and the rate constant increases.

(B) The reaction slows, and the rate constant decreases.

(C) The reaction slows, but the rate constant does not change.

(D) The reaction continues at the same rate, and the rate constant decreases.

257. The following table contains data obtained in the study of the reaction A + B → C:

Experiment	[A]$_{initial}$	[B]$_{initial}$	Initial rate (M/s)
1	0.1 M	0.1 M	4.0×10^{-2}
2	0.2 M	0.1 M	8.0×10^{-2}
3	0.2 M	0.2 M	1.6×10^{-1}

What is the rate law for this reaction?

(A) Rate = k[C] / ([A][B])

(B) Rate = k[A][B]

(C) Rate = k[A]²[B]

(D) Rate = k[A][B]²

258. The reaction of hydrogen with oxygen and the reaction of hydrogen with nitrogen are both exothermic [$\Delta H(O_2) = -241$ kJ/mol and $\Delta H(N_2) = -17$ kJ/mol]. However, the reaction of hydrogen with nitrogen is slower. Why?

(A) The activation energy is higher.

(B) The $\Delta H(N_2)$ is smaller.

(C) H_2O is more stable than NH_3.

(D) O_2 supports combustion.

259. For a particular chemical reaction, the rate constant is independent of

    (A) activation energy
    (B) enthalpy change for the reaction
    (C) temperature
    (D) presence or absence of a catalyst

260. Carbonates react with acids to release carbon dioxide gas. The rate of the reaction is directly related to the hydrogen ion concentration. Which of the following will cause the release of carbon dioxide at the fastest rate?

    (A) 0.1 M HCl
    (B) 0.1 M HF
    (C) 0.5 M H_2SO_3
    (D) 0.5 M HNO_2

261. Nitrogen oxide catalyzes the destruction of ozone in the upper atmosphere. One suggested mechanism is:

$$2\,[NO(g) + O_3(g) \rightarrow NO_2(g) + O_2(g)] \qquad \text{fast}$$
$$2\,NO_2(g) \rightarrow N_2O_4(g) \qquad \text{slow}$$
$$N_2O_4(g) + hv \rightarrow 2\,NO(g) + O_2(g) \qquad \text{fast}$$

Which of the following statements is true?

    (A) The partial pressure of NO is too low to cause a significant change in the ozone concentration.
    (B) NO lowers the activation energy of the rate-determining step.
    (C) NO has no effect on the overall energy change in the reaction.
    (D) None of the above

262. The following reaction begins very slowly but rapidly speeds up as $Mn^{2+}(aq)$ forms:

$$3\,H_2SO_4(aq) + 2\,KMnO_4(aq) + 5\,H_2C_2O_4(aq)$$
$$\rightarrow K_2SO_4(aq) + 2\,MnSO_4(aq) + 10\,CO_2(g) + 8\,H_2O(l)$$

Which of the following would change the value of the rate constant?

    (A) Adding $CO_2(g)$
    (B) Increasing the concentration of $KMnO_4(aq)$
    (C) Removing $MnSO_4(aq)$
    (D) Lowering the activation energy

263. A kinetic study of the following reaction led to a proposed mechanism:

$$Cl_2(g) + H_2O(g) \leftrightarrows HCl(g) + HClO(g)$$

The proposed mechanism was:

$Cl_2 \leftrightarrows 2\ Cl\cdot$	fast
$Cl\cdot + H_2O \rightarrow HCl + \cdot OH$	slow
$Cl\cdot + \cdot OH \rightarrow HClO$	fast

Which of the following observations would support this mechanism?

(A) The detection of $\cdot OH$
(B) The disappearance of Cl_2
(C) The determination that the rate law is Rate $= k[Cl_2]$
(D) The determination that an increase in temperature caused the reaction to go faster

264. Referring to the proposed mechanism in question 263, the first step is promoted by light. Which of the following would have the LEAST effect on the rate of the reactions in the mechanism?

(A) The addition of a substance that changes the H–O bond strength
(B) A change in the light intensity
(C) The addition of a substance that removes free radicals
(D) The addition of a small amount of argon gas

265. Which of the following statements correctly describes the two energy profiles shown here?

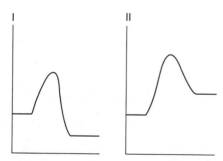

(A) Profile I is endothermic, and profile II is endothermic.
(B) Profile I is exothermic, and profile II is exothermic.
(C) Profile I is exothermic, and profile II is endothermic.
(D) Profile I is endothermic, and profile II is exothermic.

266. Which of the following energy profiles illustrates a reaction that requires energy to proceed?

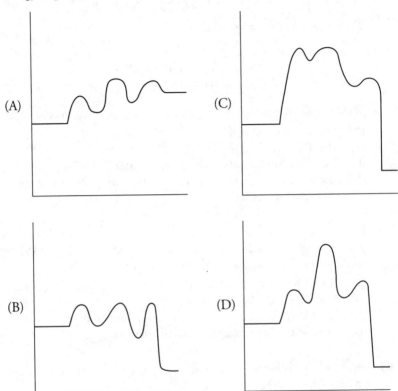

267. Which of the following energy profiles corresponds to a mechanism where the first step is the rate-determining step?

(A)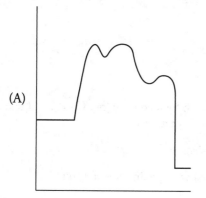

(C)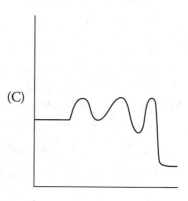

(B)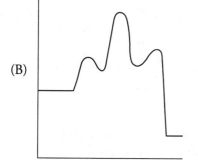

(D)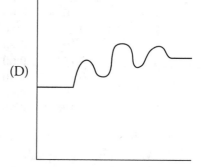

268. A certain reaction follows the rate law Rate = k[A]^2[B]. By what factor will the rate of the reaction change if both the A and B concentrations are doubled?

(A) 16
(B) 2
(C) 4
(D) 8

269. The reaction of iodide ion, I$^-$, with hypochlorite ion, ClO$^-$, will take place in basic solution. The overall reaction is:

$$I^- (aq) + ClO^- (aq) \rightarrow IO^- (aq) + Cl^- (aq)$$

The following data were obtained at a certain temperature:

Experiment	[I$^-$] (*M*)	[ClO$^-$] (*M*)	[OH$^-$] (*M*)	Relative rate of IO$^-$ formation (*M*/s)
1	0.10	0.40	0.10	2
2	0.20	0.40	0.10	4
3	0.10	0.20	0.10	1
4	0.10	0.60	0.10	8
5	0.10	0.40	0.05	4
6	0.10	0.40	0.20	1
7	0.10	0.60	0.20	1.5

What is the order of the reaction for each of the substances?

(A) [I$^-$] = +1   [ClO$^-$] = +1   [OH$^-$] = −1
(B) [I$^-$] = +1   [ClO$^-$] = +1   [OH$^-$] = 0
(C) [I$^-$] = 0   [ClO$^-$] = −1   [OH$^-$] = +1
(D) [I$^-$] = −1   [ClO$^-$] = 0   [OH$^-$] = 0

270. The reaction of iodide ion, I$^-$, with hypochlorite ion, ClO$^-$, takes place in basic solution. Referring to the same reaction and data as question 269, what is the role of the hydroxide ion?

(A) It is a catalyst.
(B) It is an inhibitor.
(C) It is a promoter.
(D) It is a scavenger.

271. The following reaction illustrates the decomposition of acetaldehyde:

$$CH_3CHO(g) \rightarrow CH_4(g) + CO(g)$$

This reaction is second order with a rate of $1.0 \times 10^{-2}\ M\,s^{-1}$ at a certain temperature. Calculate the rate constant at this temperature if the initial concentration of acetaldehyde is $0.20\ M$.

(A) $0.050\ M^{-1}\,s^{-1}$
(B) $0.10\ M^{-1}\,s^{-1}$
(C) $0.010\ M^{-1}\,s^{-1}$
(D) $0.25\ M^{-1}\,s^{-1}$

272. The following energy profile is for an exothermic reaction with $\Delta H = -45\ kJ\ mol^{-1}$ and an activation energy of $85\ kJ\ mol^{-1}$. What is the activation energy of the reverse reaction?

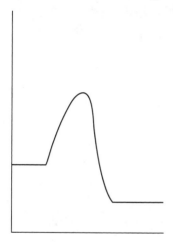

(A) $85\ kJ\ mol^{-1}$
(B) $130\ kJ\ mol^{-1}$
(C) $40\ kJ\ mol^{-1}$
(D) $45\ kJ\ mol^{-1}$

273. Consider the following reaction:

$$2 H_2O_2(aq) \rightarrow 2 H_2O(aq) + O_2(g)$$

In the absence of a catalyst, the rate of the reaction is:

$$-\frac{\Delta[H_2O_2]}{\Delta t} = 1.0 \times 10^{-8} \, M/s$$

However, in the presence of hemoglobin, which acts as a catalyst, the rate increases from 1.0×10^{-8} M/s to 2.0×10^{-1} M/s. What is the rate of oxygen production for the hemoglobin-catalyzed reaction?

(A) 1.0×10^{-1} M/s
(B) 2.0×10^{-1} M/s
(C) 1.0×10^{-8} M/s
(D) 5.0×10^{-2} M/s

274. Consider the following energy profile:

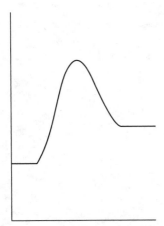

It illustrates a reaction in which

(A) the forward reaction is faster than the reverse reaction
(B) the reverse reaction is faster than the forward reaction
(C) the forward and reverse reactions proceed at the same rate
(D) it is impossible to predict the relative rates of the forward and reverse reactions

275. Which of the choices below will NOT change the rate of the following reaction?

$$CaCO_3(s) + 2\ HCl(g) \rightarrow CaCl_2(aq) + H_2O(l) + CO_2(g)$$

(A) Concentration of $CaCl_2(aq)$
(B) Partial pressure of HCl
(C) Size of $CaCO_3$ particles
(D) Temperature

276. A kinetic study of the following reaction led to a proposed mechanism:

$$Cl_2(g) + H_2O(g) \leftrightarrows HCl(g) + HClO(g)$$

The proposed mechanism was:

$Cl_2 \leftrightarrows 2\ Cl\cdot$	fast
$Cl\cdot + H_2O \rightarrow HCl + \cdot OH$	slow
$Cl\cdot + \cdot OH \rightarrow HClO$	fast

The ·OH is an example of

(A) a promoter
(B) a catalyst
(C) an inhibitor
(D) an intermediate

277. Which feature of the following energy profile would a catalyst change?

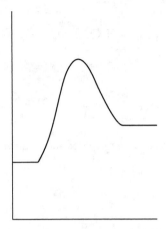

(A) The position at the end of the reaction
(B) The position at the start of the reaction
(C) The height of the peak
(D) Nothing

278. Kinetic studies of a certain reaction led to the following energy profile:

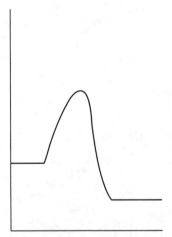

If the reaction were allowed to go to equilibrium, the value of the equilibrium constant would be

(A) >1
(B) <1
(C) =1
(D) undetermined

279. In the following energy profile, the activated complex would be at which position?

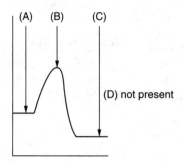

(D) not present

280. Kinetic studies on biological systems are often done at 37°C because this temperature is the normal body temperature for humans. How would the rate of reaction at 27°C compare with the rate of reaction at 37°C?

(A) One-tenth times that at 37°C
(B) Two times that at 37°C
(C) Ten times that at 37°C
(D) Half times that at 37°C

281. Which of the following reactions will proceed at the greatest rate?

(A) $A + B \rightarrow C$ $\Delta H = 100$ kJ
(B) $D + E \rightarrow F + G$ $\Delta H = 150$ kJ
(C) $H \rightarrow I + J$ $\Delta H = 250$ kJ
(D) It is impossible to determine.

282. What is the role of Br(g) in the following mechanism?

$$Br_2(g) \rightarrow 2\ Br(g) \qquad \text{slow}$$
$$2\ [Br(g) + NO(g) \rightarrow NOBr(g)] \qquad \text{fast}$$

(A) It is a Lewis base.
(B) It is a catalyst.
(C) It is a product.
(D) It is an intermediate.

283. The experimental investigation of a given reaction generated the following data:

$$2\,NO(g) + O_2(g) \rightarrow 2\,NO_2(g)$$

Experiment	[NO], M	[O$_2$], M	Rate, $\Delta[NO_2]/\Delta t$
1	0.0150	0.0150	2.40×10^{-2}
2	0.0300	0.0300	1.92×10^{-1}
3	0.0300	0.0150	9.59×10^{-2}

What is the rate law for this reaction?
(A) Rate = $k[NO]^2[O_2]$
(B) Rate = $k[NO][O_2]$
(C) Rate = $k\left(\dfrac{[NO_2]^2}{[NO]^2\,[O_2]}\right)$
(D) Rate = $k[NO_2]^2$

284. In a chemical reaction, the rate-determining step is the one that
(A) takes the least time
(B) takes the most time
(C) is the first step in the mechanism
(D) is the most complicated

285. The following table contains data obtained in the study of the reaction $A + 2\,B + C \rightarrow X + Y$:

Experiment	[A]$_{initial}$	[B]$_{initial}$	[C]$_{initial}$	Initial rate (M/s)
1	0.1 M	0.1 M	0.1 M	5.0×10^{-2}
2	0.2 M	0.1 M	0.1 M	1.0×10^{-1}
3	0.2 M	0.2 M	0.1 M	2.0×10^{-1}
4	0.1 M	0.1 M	0.3 M	4.5×10^{-1}

What is the rate law for this reaction?
(A) Rate = $k[A][B][C]^2$
(B) Rate = $k[A][B]^2[C]$
(C) Rate = $k[A][B][C]$
(D) Rate = $k[A]^2[B][C]$

286. Which of the following factors will NOT increase the value of the rate constant of a reaction?
   (A) Adding a catalyst
   (B) Increasing the temperature
   (C) Lowering the activation energy
   (D) Doubling the concentration of one of the reactants

287. For a particular chemical reaction, the rate of reaction is independent of
   (A) magnitude of the activation energy
   (B) overall enthalpy change for the reaction
   (C) absolute temperature
   (D) presence or absence of a catalyst

288. The following reaction begins very slowly but rapidly speeds up as $Mn^{2+}(aq)$ forms:

$$3 \ H_2SO_4(aq) + 2 \ KMnO_4(aq) + 5 \ H_2C_2O_4(aq)$$
$$\rightarrow K_2SO_4(aq) + 2 \ MnSO_4(aq) + 10 \ CO_2(g) + 8 \ H_2O(l)$$

One form of the rate of reaction for this overall reaction is
   (A) Rate $= - \ (½) \ \Delta[KMnO_4] \ / \ \Delta t$
   (B) Rate $= - \ (2) \ \Delta[KMnO_4] \ / \ \Delta t$
   (C) Rate $= - \ (½) \ \Delta[MnSO_4] \ / \ \Delta t$
   (D) Rate $= - \ (10) \ \Delta[CO_2] \ / \ \Delta t$

289. A kinetic study of the following reaction led to the determination of the rate law:

$$Cl_2(aq) + H_2O(l) \leftrightarrows HCl(aq) + HClO(aq)$$

Choose the *false* statement:
   (A) There is a change in the oxidation state of chlorine.
   (B) The reaction proceeds more rapidly at a higher temperature.
   (C) Increasing the partial pressure of the chlorine over the solution increases the equilibrium constant.
   (D) Both a weak and a strong acid form in the reaction.

290. A kinetic study of the following reaction led to a proposed mechanism:

$$Cl_2(g) + H_2O(g) \leftrightarrows HCl(g) + HClO(g)$$

The proposed mechanism was:

$Cl_2 \leftrightarrows 2\ Cl\cdot$	fast
$Cl\cdot + H_2O \rightarrow HCl + \cdot OH$	slow
$Cl\cdot + \cdot OH \rightarrow HClO$	fast

Based upon this mechanism, what is the rate law?

(A) Rate = $k[Cl_2]$
(B) Rate = $k[Cl_2]^{1/2}[H_2O]$
(C) Rate = $k[HCl][HClO]$
(D) Rate = $k\left(\dfrac{[Cl_2][H_2O]}{[HCl][HClO]}\right)$

291. The following reactions were studied under identical conditions:

$$(1)\ \ Ca(s) + Cl_2(g) \rightarrow CaCl_2(s)$$
$$(2)\ \ Ca(s) + I_2(g) \rightarrow CaI_2(s)$$

The rate of reaction for reaction 2 is lower than for reaction 1. What is the BEST explanation of this observation?

(A) The equilibrium constant for reaction 2 is greater than the equilibrium constant for reaction 1.
(B) The activation energy for reaction 2 is greater than the activation energy for reaction 1.
(C) $CaI_2(s)$ is more stable than $CaCl_2(s)$.
(D) $CaCl_2(s)$ is more ionic than $CaI_2(s)$.

292. Nitrogen oxide catalyzes the destruction of ozone in the upper atmosphere. One suggested mechanism is:

$$2 \, [NO(g) + O_3(g) \rightarrow NO_2(g) + O_2(g)] \qquad \text{fast}$$
$$2 \, NO_2(g) \rightarrow N_2O_4(g) \qquad\qquad\qquad\quad \text{slow}$$
$$N_2O_4(g) + hv \rightarrow 2 \, NO(g) + O_2(g) \qquad \text{fast}$$

Which of the following energy profiles applies to this mechanism?

(A)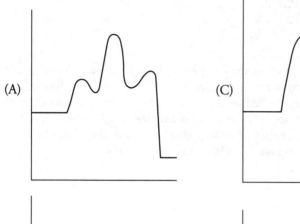

(C)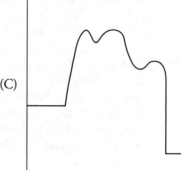

(B)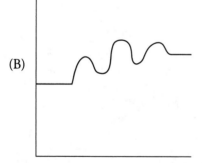

(D)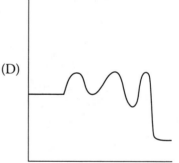

293. An experiment investigating the oxidation of iodide ion by permanganate ion in acid solution found the rate of disappearance of the permanganate ion to be 2.0×10^{-3} M/s. Determine the rate of appearance of iodine under these conditions:

$$10\ I^-(aq) + 2\ MnO_4^-(aq) + 16\ H^+(aq)$$
$$\rightarrow 2\ Mn^{2+}(aq) + 5\ I_2(aq) + 8\ H_2O(l)$$

(A) 5.0×10^{-3} M/s
(B) 2.0×10^{-3} M/s
(C) 1.0×10^{-3} M/s
(D) 2.0×10^{-2} M/s

294. Trial studies of a new drug to treat schizophrenia used an oral dose of 27 mg. The drug remained effective until the level in the bloodstream dropped to about 10% of the initial value. The drug was slowly metabolized in the body, with the metabolism following first-order kinetics with a half-life of 4 hours. How long after the initial administration of the drug would it be necessary to administer a second dose?

(A) 24 hours
(B) 4 hours
(C) 12 hours
(D) 8 hours

Equilibrium

295. The K_{sp} of AgCN is 2.2×10^{-16}, and the K_{sp} of AgOCN is 2.3×10^{-7}. Which of the following statements about these two compounds is true?

(A) The value of ΔG for the dissolution of AgCN is more positive than that of AgOCN.
(B) AgCN is more soluble than AgOCN.
(C) Both compounds are more soluble in a $0.1\ M$ AgNO$_3$ solution than in pure water.
(D) AgOCN is more soluble than AgCN.

296. The K_{sp} of AgC$_2$H$_3$O$_2$ is 4×10^{-3}. What is the minimum acetate ion concentration needed to initiate the precipitation of AgC$_2$H$_3$O$_2$ from a $1 \times 10^{-3}\ M$ Ag$^+$ solution?

(A) $4\ M$
(B) $1\ M$
(C) $4 \times 10^{-3}\ M$
(D) $1 \times 10^{-3}\ M$

297. Calcium hydroxide [Ca(OH)$_2$] is slightly soluble in water. Which of the following BEST represents the expression for the solubility product of calcium hydroxide?

(A) $K_{sp} = [\text{CaOH}^-][\text{OH}^-]$
(B) $K_{sp} = [\text{Ca(OH)}_2]$
(C) $K_{sp} = [\text{Ca}^{2+}][\text{OH}^-]^2$
(D) $K_{sp} = [\text{Ca}^{2+}][(\text{OH})^{2-}]$

298. A sample of slightly soluble calcium carbonate ($CaCO_3$) is added to water and allowed to come to equilibrium. The addition of sodium carbonate to this solution will

(A) cause more calcium carbonate to dissolve
(B) increase the calcium ion concentration
(C) lower the pH
(D) cause an increase in the amount of undissociated calcium carbonate

299. What would be the result of increasing the pressure upon the following equilibrium system?

$$CH_4(g) + H_2O(g) \leftrightharpoons CO(g) + 3\ H_2(g)$$

(A) The amount of CO would increase.
(B) The amount of H_2O would decrease.
(C) The amount of CH_4 would increase.
(D) There would be no change.

300. Which of the following statements is *false* in the study of equilibrium reactions?

(A) The presence of a catalyst does not change the value of the equilibrium constant.
(B) The addition of a reactant or product to a homogeneous gaseous equilibrium is important.
(C) The addition of a solid reactant to an equilibrium causes no change.
(D) The presence of a catalyst will increase the value of the equilibrium constant.

301. The following equilibrium is important in the industrial preparation of sulfuric acid:

$$2\ SO_2(g) + O_2(g) \leftrightharpoons 2\ SO_3(g)$$

A mixture of sulfur dioxide and oxygen is quickly introduced to a reaction chamber containing a catalyst. After the system comes to equilibrium, the partial pressure of sulfur dioxide is 10.0 atm, the partial pressure of oxygen is 5.0 atm, and the partial pressure of sulfur trioxide is 10.0 atm. What is the value of the equilibrium constant?

(A) 5.00
(B) 0.20
(C) 10.0
(D) 0.10

302. The following equilibrium is important in the industrial preparation of sulfuric acid:

$$2\,SO_2(g) + O_2(g) \leftrightarrows 2\,SO_3(g)$$

In one experiment, a mixture of sulfur dioxide and oxygen is quickly introduced to a reaction chamber. After the system comes to equilibrium, the partial pressure of the gases allows the determination of the equilibrium constant. In a second experiment, the same mixture plus a catalyst is quickly introduced to the same reaction chamber. After the system comes to equilibrium, the partial pressure of the gases allows the determination of the equilibrium constant. How do the equilibrium constants determined in the two experiments compare?

(A) They are both the same.
(B) The value for the catalyzed experiment was higher.
(C) The value for the uncatalyzed experiment was higher.
(D) It is impossible to determine from the information given.

303. Some kidney stones consist primarily of calcium oxalate. These kidney stones form only in certain patients. These patients

(A) have a low blood pH
(B) have a high blood pH
(C) also have acidosis
(D) have a calcium deficiency

304. A 0.250 g sample of insoluble AgI containing only radioactive iodine-131 is added to 1 L of water. After achieving equilibrium, 5 g of soluble KI containing nonradioactive iodine-127 is dissolved in the solution. Which of the following is true once the system reaches equilibrium a second time?

(A) The radioactive decay rate of iodine-131 increases.
(B) There will be more iodine-131 in solution.
(C) There will be the same amount of iodine-131 in solution.
(D) The solid will now be a mixture of AgI containing I-131 and AgI containing I-127.

305. The addition of $Pb(NO_3)_2$ to a solution produces a precipitate. The precipitate could be

(A) lead sulfate
(B) sodium nitrate
(C) ferric nitrate
(D) lead acetate

306. The symbol "⇆" indicates

(A) the presence of water
(B) a reversible reaction
(C) equilibrium
(D) a buffer solution

307. The autoionization of water [2 $H_2O(l)$ ⇆ $H_3O^+(aq)$ + $OH^-(aq)$] is an endothermic process. How does the pH of water change when it is heated?

(A) It decreases.
(B) It increases.
(C) It remains 7.
(D) It changes in an unpredictable manner.

308. The following equilibriums are important in solutions of H_2S:

$$H_2S(aq) ⇆ H^+(aq) + HS^-(aq) \qquad K_{a1} = 1 \times 10^{-7}$$
$$HS^-(aq) ⇆ H^+(aq) + S^{2-}(aq) \qquad K_{a1} = 1.3 \times 10^{-13}$$

The addition of sulfuric acid to a saturated solution of H_2S will NOT cause the

(A) pH to become lower
(B) concentration of undissociated H_2S to increase
(C) release of hydrogen gas
(D) K_{a2} equilibrium to shift to the left

309. The following equilibrium will form phosgene from carbon monoxide and chlorine:

$$CO(g) + Cl_2(g) ⇆ COCl_2(g)$$

What will happen to the equilibrium if the pressure on an equilibrium mixture of CO, Cl_2, and $COCl_2$ is suddenly reduced?

(A) The equilibrium will shift to the right.
(B) The equilibrium will shift to the left.
(C) The equilibrium will not shift.
(D) It is impossible to determine what will happen to the equilibrium.

310. The following exothermic reaction will form phosgene from carbon monoxide and chlorine:

$$CO(g) + Cl_2(g) \leftrightarrows COCl_2(g)$$

An increase in temperature will cause

(A) an unknown change
(B) the equilibrium to shift to the right
(C) no change in the equilibrium
(D) the equilibrium to shift to the left

311. The content of air exiting the lungs during normal breathing is

(A) higher in carbon dioxide and lower in oxygen relative to air
(B) higher in carbon dioxide and higher in oxygen relative to air
(C) lower in carbon dioxide and lower in oxygen relative to air
(D) lower in carbon dioxide and higher in oxygen relative to air

312. A mixture is placed in a cylinder with a moving piston. The mixture achieves equilibrium and then the piston moves in, reducing the volume. Which of the following equilibriums will undergo the greatest change?

(A) $Ni(s) + 4\,CO(g) \leftrightarrows Ni(CO)_4(g)$
(B) $CaCO_3(s) \leftrightarrows CaO(s) + CO_2(g)$
(C) $Fe^{3+}(aq) + 6\,CN^-(aq) \leftrightarrows [Fe(CN)_6]^{3-}(aq)$
(D) $2\,H_2(g) + O_2(g) \leftrightarrows 2\,H_2O(g)$

313. The following reaction is at equilibrium:

$$Cl_2(aq) + H_2O(l) \leftrightarrows HCl(aq) + HClO(aq)$$

A quantity of $HCl(g)$ dissolves in the mixture and triples the concentration of HCl. What effect will the HCl have on the equilibrium?

(A) The reaction quotient will be one-third the equilibrium constant.
(B) The reaction quotient will be triple the equilibrium constant.
(C) The reaction quotient will be equal to the equilibrium constant.
(D) The reaction quotient will be equal to 3.

314. Some chlorine oxides will dissolve in water to produce acids. Two examples are:

$$Cl_2O(g) + H_2O(l) \leftrightharpoons 2\ HClO(aq)$$
$$Cl_2O_5(g) + H_2O(l) \leftrightharpoons 2\ HClO_3(aq)$$

Why is Cl_2O_5 more soluble than Cl_2O?

(A) $HClO_3$ is a strong acid, and its dissociation shifts its equilibrium to the right.

(B) $HClO_3$ is a strong acid, and its dissociation shifts its equilibrium to the left.

(C) HClO is a strong acid, and its dissociation shifts its equilibrium to the right.

(D) HClO is a strong acid, and its dissociation shifts its equilibrium to the left.

315. The solubility constant of calcium fluoride, CaF_2, is 4.0×10^{-11}. Will calcium fluoride precipitate from a solution prepared by adding 10.0 mL of 0.01 M $CaCl_2$ and 20.0 mL of 0.01 M HF to sufficient water to make 1.0 L of solution?

(A) Yes

(B) No

(C) Yes, if the temperature is increased

(D) There is insufficient data.

316. Which of the following would be most effective at decreasing the solubility of AgCl in water?

(A) 1 M HCl

(B) 1 M NaCl

(C) 1 M $CaCl_2$

(D) 1 M CCl_3COOH

317. Which of the choices below would have a minimal effect on the following equilibrium?

$$CaCO_3(s) \leftrightharpoons CaO(s) + CO_2(g)$$

(A) Adding CaO to the container

(B) Adding CO_2 to the container

(C) Heating the mixture in the container

(D) Decreasing the volume of the container

318. The following equilibrium occurs at a certain temperature:

$$CH_4(g) + H_2O(g) \leftrightarrows CO(g) + 3\ H_2(g)$$

Which of the following would have a minimal effect on the equilibrium concentrations?

(A) Changing the temperature
(B) Removing CO
(C) Adding CH_4
(D) Adding a catalyst

319. The following equilibrium occurs at a certain temperature:

$$N_2(g) + O_2(g) \leftrightarrows 2\ NO(g)$$

Which of the following would have a minimal effect on the equilibrium concentrations?

(A) Changing the temperature
(B) Changing the pressure
(C) Changing the amount of NO
(D) Changing the amount of N_2

320. The following equilibrium occurs at a certain temperature:

$$N_2(g) + O_2(g) \leftrightarrows 2\ NO(g)$$

What is the value of the equilibrium constant of this reaction if the equilibrium concentrations are $[N_2] = 0.10\ M$, $[O_2] = 0.20\ M$, and $[NO] = 0.20\ M$?

(A) 10
(B) 2.0
(C) 0.10
(D) 0.50

321. A solution of AgCN ($K_{sp} = 2.2 \times 10^{-16}$) is at equilibrium. The addition of a small quantity of NaCN will

(A) cause more AgCN to dissolve
(B) cause more AgCN to precipitate
(C) increase the value of the solubility constant
(D) decrease the value of the equilibrium constant

322. The solubility of AgCN in water is about 1.0×10^{-8} mol/L. What is the solubility product constant for AgCN?

(A) 1.0×10^{-24}
(B) 1.0×10^{-8}
(C) 1.0×10^{-16}
(D) 2.0×10^{-8}

323. $CaSO_4$ is added to a 0.1 M Na_2SO_4 solution. The solubility of $CaSO_4$ is

(A) lowered, due to the common ion effect
(B) increased, due to the common ion effect
(C) unchanged, because Na^+ compounds are always soluble
(D) lowered, because the Na_2SO_4 combines with most of the water, leaving little to dissolve the $CaSO_4$

324. What is the BEST way of expressing the equilibrium constant for the following reaction?

$$2 \, NO(g) + O_2(g) \leftrightarrows 2 \, NO_2(g)$$

(A) $[NO_2]^2 / [NO]^2[O_2]$
(B) $2[NO_2] / 2[NO][O_2]$
(C) $[NO]^2[O_2] / [NO_2]^2$
(D) $[NO_2] / [NO][O_2]$

325. Which of the choices below would increase the production of $H_2(g)$ in the following equilibrium?

$$CH_4(g) + H_2O(g) \leftrightarrows CO(g) + 3 \, H_2(g)$$

(A) Adding CO
(B) Removing H_2
(C) Removing CH_4
(D) Increasing the pressure

326. Two experiments were used to study the following equilibrium:

$$CH_4(g) + H_2O(g) \leftrightarrows CO(g) + 3\,H_2(g)$$

Both experiments began with identical quantities of CH_4 and H_2O in identical containers at the same temperature. However, before beginning the second experiment, a catalyst was added to the system. Both systems go to equilibrium. At the end of each experiment

(A) the concentrations of all reactants and products are the same in each system
(B) the concentrations of the products are higher in the system containing the catalyst
(C) the concentrations of the reactants are higher in the system without a catalyst
(D) it is not possible to determine the relative amounts of reactants and products

327. The reaction of nitrogen gas with hydrogen, in the presence of a catalyst, will reach an equilibrium with $\Delta H = -17$ kJ/mol NH_3. The equilibrium reaction is:

$$N_2(g) + 3\,H_2(g) \leftrightarrows 2\,NH_3(g)$$

Increasing the temperature of this system in equilibrium will

(A) increase the kinetic energy of the molecules
(B) increase the concentration of ammonia
(C) not change anything
(D) decrease the total pressure

328. The following equilibrium is important in the industrial preparation of sulfuric acid:

$$2\,SO_2(g) + O_2(g) \leftrightarrows 2\,SO_3(g)$$

A two-to-one mixture of sulfur dioxide and oxygen is quickly introduced to a reaction chamber containing a catalyst. The initial pressure of the mixture is 30.0 atm. After the system comes to equilibrium, the total pressure is 25.0 atm. What is the partial pressure of sulfur trioxide at equilibrium?

(A) 5.0 atm
(B) 10.0 atm
(C) 25.0 atm
(D) 7.5 atm

329. Calcium fluoride, CaF_2, has a solubility constant of 4.0×10^{-11}. Calcium fluoride will be LEAST soluble in which of the following?

(A) 0.1 M $Ba(NO_3)_2$
(B) Pure water
(C) 0.1 M $NaNO_3$
(D) 0.1 M $Ca(NO_3)_2$

330. The solubility product constant for cupric arsenate is 10^{-37}. What is the approximate arsenate concentration in a saturated solution of cupric arsenate?

(A) 10^{-37} M
(B) 10^{-8} M
(C) 10^{-7} M
(D) 10^{-18} M

331. A 0.250 g sample of insoluble AgI containing only radioactive iodine-131 is added to 1 L of water. After achieving equilibrium, 5 g of soluble KI containing nonradioactive iodine-127 is dissolved in the solution. Which of the following is true once the system reaches equilibrium a second time?

(A) The radioactive decay rate of iodine-131 increases.
(B) There will be less iodine-131 in solution.
(C) There will be the same amount of iodine-131 in solution.
(D) There will be more iodine-131 in solution.

332. What is the concentration of a solution prepared by adding 300 g of calcium carbonate to 500 mL of water?

(A) 3 M
(B) 6 M
(C) 0.6 M
(D) <0.6 M

CHAPTER 9

Acids and Bases

333. The solution formed by the dissolution of an oxide of manganese in water had a pH of 1.0. Which of the following oxides of manganese might be present?

(A) MnO_2
(B) MnO
(C) Mn_2O_3
(D) Mn_2O_7

334. The hydrogen phosphate ion (HPO_4^{2-}) can serve as an acid. Which of the following will result from the acidic dissociation of the hydrogen phosphate ion?

(A) PO_4^{3-}
(B) PO_4^{2-}
(C) $H_2PO_4^{-}$
(D) H_3PO_4

335. Which of the following can combine with HNO_2 to form a buffer solution?

(A) HCl
(B) KOH
(C) $NaNO_3$
(D) C_2H_5OH

336. All the hydrogen halides behave as acids in solution. Which of the following might be the conjugate base of a hydrogen halide?

(A) F^-
(B) F
(C) OH^-
(D) H^+

337. Which of the following procedures would indicate the presence of carbonate ion in a solution?

(A) The addition of a base to cause the generation of a gas
(B) The addition of an acid to cause the generation of a gas
(C) The addition of NaCl to form a precipitate
(D) The addition of $(NH_4)_2SO_4$ to form a precipitate

338. Which of the following acids is the strongest?

(A) CCl_3COOH
(B) $CHCl_2COOH$
(C) $CH_2ClCOOH$
(D) CH_3COOH

339. The pH of a 10^{-3} M solution of HI is

(A) <3
(B) 3
(C) >3
(D) unknown

340. The pH of blood is about 7, and the pH of stomach acid is about 1. The hydrogen ion concentration in stomach acid is how many times greater than that of blood?

(A) 10^6
(B) 10^{-6}
(C) 6
(D) −6

341. Ten milliliters of a hydrochloric acid solution with a pH of 1 is diluted to 100 mL. The pH of the diluted solution is

(A) 1
(B) 2
(C) 0.1
(D) 7

342. Water from abandoned lead mines is often acidic. Which of the following substances might cause the water to be acidic?

(A) SO_2

(B) SiO_2

(C) $CaCO_3$

(D) PbS

343. Which of the following compounds can buffer a solution against both acids and bases?

(A) Na_2CO_3

(B) $NaHCO_3$

(C) NH_4Cl

(D) $CaCl_2$

344. The pH of blood is about 7.4. Which of the following is NOT true concerning blood?

(A) $CaCO_3$ is more soluble in blood than in water.

(B) The hydrogen ion concentration is less than the hydroxide ion concentration.

(C) The pOH is about 6.6.

(D) Organic acids will be more soluble in blood than in pure water.

345. An acid-base indicator may use any of the following functional groups EXCEPT

(A) sulfonic acid

(B) amine

(C) carboxylic acid

(D) alcohol

346. The K_a of the methylammonium ion ($CH_3NH_3^+$) is 2×10^{-12}. What is the approximate pH of a 1 M solution of methylammonium chloride?

(A) 6

(B) 3

(C) 12

(D) 8

347. Weak acids, such as formic acid (CHOOH), exist in equilibrium in solution. The equilibrium for formic acid is:

$$CHOOH(aq) \leftrightharpoons H^+(aq) + CHCOO^-(aq)$$

The addition of potassium hydroxide to a formic acid solution will cause all of the following EXCEPT

(A) an increase in the ability of the solution to conduct electricity
(B) a shift in the equilibrium to the right
(C) a decrease in the pH
(D) an increase in the concentration of the formate ion

348. The addition of solid potassium hydroxide to a 0.1 M nitric acid solution results in the neutralization of 99% of the acid. What is the pH of the solution?

(A) 7
(B) 1
(C) 3
(D) 14

349. In order to maximize the buffer capacity of a solution it is BEST to

(A) maximize the concentrations of the conjugate acid and conjugate base
(B) minimize the concentrations of the conjugate acid and conjugate base
(C) adjust the pH to near 7
(D) choose a system where the pK_a is the same as the pK_b

350. Which of the following is an example of a polyprotic base?

(A) CaO
(B) PO(OH)$_3$
(C) NH$_3$
(D) H$_2$SO$_4$

351. The process producing water from H^+ and OH^- is very exothermic. This means that when heating water, the value of K_w

(A) will change, but it is necessary to know the pressure to know how
(B) decreases
(C) remains constant
(D) increases

352. A water sample was contaminated with a sodium halide. The pH of the water sample was greater than 7. Which sodium halide is the most likely contaminant?

(A) NaF
(B) NaCl
(C) NaBr
(D) NaI

353. At room temperature, the pH of household ammonia is 11. The pH of vinegar is 3. The $[H^+]$ in vinegar is how much higher than in household ammonia?

(A) ⅛
(B) 8
(C) 10^8
(D) 800

354. The K_a (ionization constant) value for a strong acid is

(A) $K_a = 1$
(B) $K_a < 1$
(C) $K_a > 1$
(D) K_a depends on the identity of the strong acid.

355. Arsenic acid, H_3AsO_4, is a weak triprotic acid. Which of the following is true concerning the concentrations in a 0.10 M arsenic acid solution?

(A) $[AsO_4^{3-}] > [HAsO_4^{2-}]$
(B) $[H^+] > [H_2AsO_4^-]$
(C) $[H^+] < [H_2AsO_4^-]$
(D) $[AsO_4^{3-}] = [H_3AsO_4]$

356. A solution was prepared by dissolving 1.0 mole of a compound in 500.0 mL of water. Later the solution was diluted to 1.500 L. The pH of the resultant solution was 4.3. What is the relationship between the K_a of the compound and the K_b of its conjugate base?

(A) $K_a < K_b$
(B) $K_a > K_b$
(C) $K_a = K_b$
(D) $K_a = 1 / K_b$

357. The following bond energies and K_{a1} values were measured for compounds with the general formula H_2X, where X is an element in the oxygen family:

Bond	Bond energy (kJ mol^{-1})	K_{a1}
H$-$O	428	—
H$-$S	344	1.0×10^{-7}
H$-$Se	305	1.3×10^{-4}
H$-$Te	268	2.3×10^{-3}

What is the relationship between the bond energies in the table, the electronegativity, and the K_a values for the H_2X compounds?

(A) The acidity increases as the bond energy and electronegativity decrease.

(B) The acidity increases as the bond energy decreases and electronegativity increases.

(C) The acidity increases as the bond energy increases and electronegativity decreases.

(D) The acidity decreases as the bond energy and electronegativity decrease.

358. Which of the following BEST represents the acidity of the hydrogen halides?

(A) HF = HCl > HBr = HI

(B) HF > HCl > HBr > HI

(C) HI > HBr > HCl > HF

(D) HI > HBr > HCl = HF

359. Consider the following K_a values:

Acid	K_a
HClO (hypochlorous acid)	3.5×10^{-8}
HC$_3$H$_3$O$_3$ (pyruvic acid)	1.4×10^{-4}

Which of the following would be protonated to the greater extent by the addition of 10.0 mL to a 0.10 M solution of ClO$^-$ or C$_3$H$_3$O$_3^-$?

(A) ClO$^-$, because it is a stronger base than C$_3$H$_3$O$_3^-$

(B) ClO$^-$, because it is a weaker base than C$_3$H$_3$O$_3^-$

(C) C$_3$H$_3$O$_3^-$, because it is a stronger base than ClO$^-$

(D) C$_3$H$_3$O$_3^-$, because it is a weaker base than ClO$^-$

360. When dissolved in water, strong acids appear equal in strength. However, differences in strength are noted in other solvents. Which would be the BEST solvent to differentiate two strong acids?

(A) $NaOH(aq)$
(B) $NH_3(l)$
(C) $(CH_3)_3N(l)$
(D) $HC_2H_3O_2(l)$

361. Amino acids, the constituents of proteins, are molecules with a central carbon atom attached to an amine group, a carboxylic acid group, hydrogen, and an R group. The identity of the R group identifies the amino acid. The following equilibria are important to all amino acids:

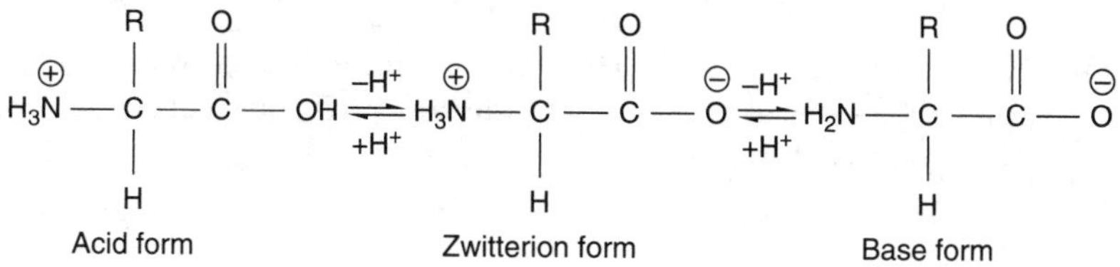

Acid form   Zwitterion form   Base form

Which of the following BEST describes the behavior of amino acids?

(A) They are weak acids.
(B) They are amphoteric.
(C) They are weak bases.
(D) They are ionizable.

362. The reaction of 25.00 mL of 1.5 M NaOH with 25.00 mL of 1.5 M HNO$_3$ in a calorimeter resulted in a 15.0°C increase in the temperature. What would be the pH of the solution after the mixture cooled to the original temperature?

(A) >7
(B) <7
(C) 7
(D) Impossible to determine

363. Identify the Brønsted-Lowry base in the following reaction:

$$HNO_3(aq) + H_2SO_4(aq) \rightarrow (NO_2^+)(HSO_4^-)(aq) + H_2O(l)$$

(A) $HNO_3(aq)$
(B) $H_2SO_4(aq)$
(C) $(NO_2^+)(HSO_4^-)(aq)$
(D) $H_2O(l)$

364. The K_a of HOCN is 3.5×10^{-4}. What is the approximate concentration of OCN$^-$ ions in a 1 M solution of HOCN?

    (A) $3.5 \times 10^{-4}\ M$
    (B) $1.9 \times 10^{-2}\ M$
    (C) $1.5 \times 10^{-8}\ M$
    (D) 1 M

365. The addition of sodium cyanate to a cyanic acid solution causes the pH to

    (A) increase, due to the common ion effect
    (B) decrease, due to the common ion effect
    (C) remain constant, because a buffer forms
    (D) remain constant, because sodium cyanate is a neutral salt

366. An acid with a K_a of 2×10^{-6} is used to make a buffer. What is the approximate pH of the solution if the buffer has equal concentrations of both the acid and its conjugate base?

    (A) Between 4 and 5
    (B) Between 2 and 3
    (C) Between 3 and 4
    (D) Between 5 and 6

367. What is the pH of a 0.010 M HNO$_3$ solution?

    (A) 1.0
    (B) 0.010
    (C) 2.0
    (D) 0.0

368. Which of the following is NOT a weak base?

    (A) CH_3NH_2
    (B) NH_3
    (C) KOH
    (D) F$^-$

369. Which of the following solutes will produce a basic solution when dissolved in water?

    (A) C_2H_5OH
    (B) H_3PO_4
    (C) NaCl
    (D) Na_2CO_3

370. Which of the following statements is NOT true?

(A) HCO_3^- is the conjugate base of H_2CO_3.

(B) CH_3COOH is the conjugate acid of CH_3COO^-.

(C) HNO_3 is the conjugate acid of NO_2^-.

(D) NH_3 is the conjugate base of NH_4^+.

371. The following equilibria are important in solutions of H_2S:

$$H_2S(aq) \leftrightarrows H^+(aq) + HS^-(aq) \qquad K_{a1} = 1 \times 10^{-7}$$
$$HS^-(aq) \leftrightarrows H^+(aq) + S^{2-}(aq) \qquad K_{a2} = 1.3 \times 10^{-13}$$

The pK_a of H_2S is about

(A) 20

(B) 13

(C) 7

(D) 1

372. Which of the following compounds will increase the hydroxide ion concentration when dissolved in water?

(A) NH_4Cl

(B) CH_3CH_2OH

(C) K_3PO_4

(D) $NaCl$

373. Which of the following indicators would be the BEST one to use in the acid-base titration when adding a strong base from a burette to a weak acid?

Indicator	Color change	pH range
thymol blue	yellow-blue	7.4–9.2
nitrazine yellow	yellow-blue-violet	6.0–7.0
methyl red	red-yellow	4.2–6.2
methyl violet	green-blue	1.0–1.5

(A) Nitrazine yellow

(B) Thymol blue

(C) Methyl red

(D) Methyl violet

374. The K_a of the methylammonium ion ($CH_3NH_3^+$) is 2×10^{-12}. What is the approximate pH of a solution with a methylamine concentration equal to the methylammonium ion concentration of the same solution?

    (A) 6

    (B) 3

    (C) 12

    (D) 8

375. Which of the following combinations will yield a buffer solution?

    (A) $NaHCO_3$ and Na_2CO_3

    (B) HCl and NaCl

    (C) CaO and $Ca(OH)_2$

    (D) $C_2H_5OH_2^+Cl^-$ and C_2H_5OH

376. A solution prepared by mixing equal volumes of $0.50\ M$ nitric acid with $0.50\ M$ ammonia would have

    (A) the same pH as a $0.25\ M$ ammonium nitrate solution

    (B) the same pH as $0.50\ M$ ammonium nitrate

    (C) a pH above 7

    (D) the same pH as $1.0\ M$ ammonium nitrate

377. Methemoglobin forms when the Fe^{2+} undergoes oxidation to Fe^{3+}. The nitrite ion can cause this oxidation. During nitrite oxidation, the nitrogen in the nitrite ion undergoes reduction from $3+$ to 0. What is the oxidation state of iron in hemoglobin?

    (A) $+3$

    (B) $+2$

    (C) The same as that of nitrogen in the nitrite ion

    (D) Impossible to determine

378. Consider the following K_a values:

Acid	K_a
$HC_2H_3O_2$ (acetic acid)	1.7×10^{-5}
HCN (hydrocyanic acid)	4.9×10^{-10}
$HClO$ (hypochlorous acid)	3.5×10^{-8}
$HC_3H_3O_3$ (pyruvic acid)	1.4×10^{-4}

Which of the following is the strongest base?

(A) CN^-
(B) $C_2H_3O_2^-$
(C) ClO^-
(D) $C_3H_3O_3^-$

CHAPTER 10

Thermodynamics

379. The following reaction will form phosgene from carbon monoxide and chlorine:

$$CO(g) + Cl_2(g) \rightarrow COCl_2(g)$$

During this reaction

(A) there is a decrease in the entropy of the system
(B) there is an increase in the entropy of the system
(C) the entropy of the system remains constant
(D) the change in entropy is impossible to predict

380. The following reaction will form phosgene from carbon monoxide and chlorine:

$$CO(g) + Cl_2(g) \rightarrow COCl_2(g)$$

Consider the following thermodynamic data:

Compound	ΔH_f° (kJ/mol)	ΔG_f° (kJ/mol)	S° (J/mol·K)
$CO(g)$	-110	-137	198
$Cl_2(g)$	0	0	223
$COCl_2(g)$	-223	-210	290

Which of the following will most likely change the ΔG° for the reaction?

(A) Increasing the reaction rate
(B) Adding a catalyst
(C) Lowering the activation energy
(D) Changing the temperature

381. Under a specific set of conditions, a reaction is not spontaneous. What must be true about this reaction under these conditions?

(A) ΔH is positive.
(B) ΔS is negative.
(C) ΔG is positive.
(D) $\Delta G = 0$

382. Which of the following statements is NOT true?

(A) The entropy of gases is usually greater than that of a solid.
(B) Cooling any substance to absolute zero reduces the entropy of the substance to zero.
(C) Under standard conditions, ΔH_f° and ΔG_f° for any element in its standard state is zero.
(D) STP and standard conditions are not the same.

383. The standard heat of formation of $HF(g)$ is -270 $kJ \cdot mol^{-1}$. Based upon this information, what is the ΔH_{rxn} for the following reaction under standard conditions?

$$H_2(g) + F_2(g) \rightarrow 2\ HF(g)$$

(A) -540 $kJ \cdot mol^{-1}$
(B) -270 $kJ \cdot mol^{-1}$
(C) $+540$ $kJ \cdot mol^{-1}$
(D) More information is necessary.

384. Natural gas is primarily methane. The combustion of natural gas is important in heating homes because it is a very exothermic process. This means that K_p for the combustion of methane is

(A) negative
(B) positive and small
(C) positive and large
(D) zero

385. A sample of hydrogen gas is in an adiabatic system. This means that

(A) hydrogen will not react with the sides of the container
(B) the pressure remains constant
(C) the temperature remains constant
(D) there is no heat exchange between the system and the surroundings

386. An inexpensive substitute for a calorimeter is an insulated foam coffee cup. A major limitation of using a foam coffee cup for a calorimeter is that it will not work effectively under which of the following conditions?

(A) At very high temperatures
(B) At very high pressure
(C) At very low pressure
(D) All of the above

387. The reaction of 25.00 mL of 1.5 M NaOH with 25.00 mL of 1.5 M HNO$_3$ in a calorimeter resulted in a 15.0°C increase in the temperature. Repeating the experiment with 25.00 mL of 1.5 M acetic acid replacing the HNO$_3$ would

(A) result in a smaller temperature change, because the enthalpy change is smaller
(B) result in a larger temperature change, because the enthalpy change is larger
(C) result in the same temperature change, because all acid-base reactions have the same enthalpy change
(D) not be predictable based on the information given

388. The following reaction occurred in a system at constant temperature:

$$Ca(s) + Cl_2(g) \rightarrow CaCl_2(s)$$

How does the entropy of the system change as the reaction progresses?

(A) It depends on the temperature.
(B) It increases.
(C) It remains the same.
(D) It decreases.

389. The following reaction is at equilibrium:

$$SnCl_2(s) + Cl_2(g) \leftrightarrows SnCl_4(l)$$

A chemist then adds 0.50 mole of SnCl$_2$. In which direction will the reaction proceed?

(A) To the left
(B) To the right
(C) No change
(D) Impossible to determine from the information given

390. Which of the following combinations indicates a system with a process that is always spontaneous?

(A) ΔH is negative, and ΔS is negative.
(B) ΔH is negative, and ΔS is positive.
(C) ΔH is positive, and ΔS is negative.
(D) ΔH is positive, and ΔS is positive.

391. Which of the following combinations indicates a system with a process that requires heating to become spontaneous?

(A) ΔH is positive, and ΔS is negative.
(B) ΔH is negative, and ΔS is negative.
(C) ΔH is negative, and ΔS is positive.
(D) ΔH is positive, and ΔS is positive.

392. What is the change in the internal energy when a sample of argon gas is compressed? The process required 1,475 J of work done on the gas and the removal of 375 J of heat energy.

(A) $-1,100$ J
(B) 1,850 J
(C) 1,100 J
(D) $-1,850$ J

393. When a liquid freezes, there is a negative entropy change in the system. The second law of thermodynamics states that for any spontaneous process, the entropy change must be positive. How can a liquid freeze spontaneously when the entropy change of the system is negative?

(A) The second law of thermodynamics does not apply to phase changes.
(B) The removal of heat during freezing causes a greater negative enthalpy change in the surroundings.
(C) There are more microstates available in the solid; therefore, the solid allows more options.
(D) The first law of thermodynamics supersedes the second law, because freezing also involves the transfer of energy.

394. Methyl alcohol evaporates from a glass. Indicate whether the expected signs of ΔS and ΔH are to be positive or negative.

(A) ΔH is positive, and ΔS is positive.
(B) ΔH is negative, and ΔS is negative.
(C) ΔH is negative, and ΔS is positive.
(D) ΔH is positive, and ΔS is negative.

395. The enthalpy change for the vaporization of CCl_4 is positive. Why is the value positive?

(A) Energy is required to break the dipole-dipole forces present.
(B) Energy is required to break the carbon-chlorine covalent bonds.
(C) Energy is required to break the London dispersion forces present.
(D) Energy is required to break the carbon-chlorine ionic bonds.

396. The Gibbs free energy change for a process is positive. What can be said about the equilibrium constant for the process?

(A) $K > 1$
(B) $K < 1$
(C) $K = 1$
(D) Nothing, as the two are unrelated

397. Calculate the work involved in the following reaction when 4.0 g of calcium reacts with excess water to generate hydrogen gas at 0.0°C and 1.00 atm (1 J = 100 L·atm):

$$Ca(s) + 2\ H_2O(l) \rightarrow Ca(OH)_2(aq) + H_2(g)$$

(A) −220 J
(B) +220 J
(C) +100 J
(D) −100 J

398. The following is the energy profile for a reaction:

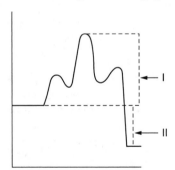

The overall enthalpy change for the reaction is

(A) I
(B) II
(C) I + II
(D) II − I

399. The following reaction will form phosgene from carbon monoxide and chlorine:

$$CO(g) + Cl_2(g) \rightarrow COCl_2(g)$$

Consider the following thermodynamic data:

Compound	ΔH_f° (kJ/mol)	ΔG_f° (kJ/mol)	S° (J/mol·K)
$CO(g)$	-110	-137	198
$Cl_2(g)$	0	0	223
$COCl_2(g)$	-223	-210	290

The enthalpy change for the formation of 1 mole of phosgene is
(A) -131 kJ
(B) -223 kJ
(C) -210 kJ
(D) -113 kJ

400. To guarantee that a reaction has a positive free energy change, the enthalpy change must be
(A) negative and the entropy change must be negative
(B) positive and the entropy change must be negative
(C) positive and the entropy change must be positive
(D) negative and the entropy change must be positive

401. Due to the greenhouse effect, a sample of gas (the system) absorbs more energy than it emits. This means that q is
(A) positive
(B) negative
(C) equal to ΔE + work
(D) Both B and C

402. Which of the following reactions will probably have the greatest increase in ΔS_{rxn}?
(A) $C_6H_{12}O_6(s) + 6\,O_2(g) \rightarrow 6\,CO_2(g) + 6\,H_2O(g)$
(B) $H_2O(l) \rightarrow H_2O(g)$
(C) $6\,CO_2(g) + 6\,H_2O(g) \rightarrow C_6H_{12}O_6(s) + 6\,O_2(g)$
(D) $CH_4(g) + 2\,O_2(g) \rightarrow CO_2(g) + 2\,H_2O(g)$

403. Heating a 150.0 g sample of water from 25°C to 75°C results in an increase in the entropy of the water. The main cause for this increase in entropy is that

(A) the average kinetic energy of the water molecules increases
(B) most of the hydrogen bonds break
(C) the water vaporizes
(D) the viscosity of the water increases

404. An inexpensive substitute for a calorimeter is an insulated foam coffee cup. The major reason why a coffee cup is an effective substitute is that it

(A) is nearly adiabatic
(B) remains isothermal
(C) is lightweight
(D) is unreactive

405. The following high-temperature reaction is useful in the production of elemental barium:

$$4\ BaO(s) + Si(s) \rightarrow Ba_2SiO_4(s) + 2\ Ba(g)$$

Which of the following statements is true about this reaction?

(A) Silicon undergoes oxidation.
(B) The reaction is spontaneous at room temperature.
(C) There is a decrease in entropy.
(D) Ionic BaO changes to molecular Ba_2SiO_4.

406. The following equation represents the reaction of methane gas with a limited amount of oxygen gas:

$$2\ CH_4(g) + 3\ O_2(g) \rightarrow 2\ CO(g) + 4\ H_2O(g)$$

What are the signs of ΔH and ΔS for this reaction?

(A) ΔH is positive, and ΔS is negative.
(B) ΔH is negative, and ΔS is negative.
(C) ΔH is positive, and ΔS is positive.
(D) ΔH is negative, and ΔS is positive.

407. The following data came from three experiments using a calorimeter:

Experiment 1   Dissolving 8.0 g of NaOH(s) in 100.0 mL of water resulted in a 15°C temperature increase.

Experiment 2   Mixing 50.00 mL of 4.00 M NaOH with 50.00 mL of 4.00 M HNO_3 resulted in a 52°C temperature increase.

Experiment 3   Dissolving 8.0 g of NaOH(s) in 100.0 mL of 2.00 M HNO_3 resulted in a 65°C temperature increase.

Why are the temperature changes in Experiment 2 and Experiment 3 not the same when both experiments used equal numbers of moles and had the same final volume?

(A) The molarities of the HNO_3 solutions were different.
(B) When NaOH(s) dissolves, the process is endothermic.
(C) When NaOH(s) dissolves, the process is exothermic.
(D) The reaction in Experiment 2 did not go to completion.

408. A student builds the apparatus shown here:

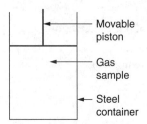

For this system, $\Delta E = q + w$, where q is heat and w is work. (Heat lost by the system is negative, while heat gained by the system is positive.) In an experiment, the student began with a 0.50 mole sample of argon gas at 0°C and 1.00 atm. The sample was heated at constant pressure to 273°C, the final volume determined, and the piston locked into place. Next, the system is wrapped with insulation to prevent any heat exchange with the surroundings. Finally, the piston is unlocked. Which of the following is true?

(A) q > 0
(B) q = 0
(C) q < 0
(D) $\Delta T = 0$

409. The following reaction is at equilibrium:

$$SnCl_2(s) + Cl_2(g) \leftrightarrows SnCl_4(l)$$

The standard Gibbs energy of formation for the materials in the reaction is as follows:

$SnCl_2(s)$ −342 kJ/mol
$Cl_2(g)$ 0.0 kJ/mol
$SnCl_4(l)$ −440 kJ/mol

What is the value of ΔG?

(A) 0 kJ/mol
(B) −98 kJ/mol
(C) +98 kJ/mol
(D) Impossible to determine from the information given

410. A chemist studies the following reaction:

$$SnCl_2(s) + Cl_2(g) \leftrightarrows SnCl_4(l)$$

The standard enthalpies of formation for the materials in the reaction are:

$SnCl_2(s)$ −325 kJ/mol
$Cl_2(g)$ 0.0 kJ/mol
$SnCl_4(l)$ −511 kJ/mol

What is the value of ΔH for the reaction?

(A) +186 kJ
(B) −186 kJ
(C) 0 kJ
(D) −836 kJ

411. For the melting of aluminum, $Al(s) \leftrightarrows Al(l)$, the heat of fusion is 10.0 kJ/mol and the entropy of fusion is 9.50 J/mol·K. Estimate the melting point of Al.

(A) 1,000 K
(B) 730 K
(C) 1,270 K
(D) There is insufficient information.

CHAPTER 11

Electrochemistry

412. The compound $CuCl_2$ is most likely

(A) a reducing agent
(B) an oxidizing agent
(C) insoluble in water
(D) a nonelectrolyte

413. The following reaction has a standard cell potential of +1.66 V:

$$2 \text{ Al}(s) + 6 \text{ H}^+(aq) \rightarrow 2 \text{ Al}^{3+}(aq) + 3 \text{ H}_2(g)$$

The value of +1.66 V means

(A) decreasing the amount of Al(s) will increase the cell potential
(B) the reverse reaction is spontaneous
(C) increasing the amount of Al(s) will increase the cell potential
(D) the reverse reaction is nonspontaneous

414. Which of the following will show a variation of cell potential with pressure?

(1) $Mg(s) + 2 \text{ H}^+(aq) \rightarrow Mg^{2+}(aq) + H_2(g)$
(2) $Mg(s) + Cu^{2+}(aq) \rightarrow Mg^{2+}(aq) + Cu(s)$
(3) $H_2(g) + 2 \text{ Ag}^+(aq) \rightarrow 2 \text{ H}^+(aq) + 2 \text{ Ag}(s)$
(4) $14 \text{ H}^+(aq) + 6 \text{ Fe}^{2+}(aq) + Cr_2O_7^{2-}(aq)$
$\rightarrow 6 \text{ Fe}^{3+}(aq) + 2 \text{ Cr}^{3+}(aq) + 7 \text{ H}_2O(l)$

(A) 1 and 3
(B) 2 and 4
(C) 1 and 2
(D) 3 and 4

415. According to the Nernst equation, what will happen to the cell potential of the following reaction after doubling the quantity of solid silver?

$$H_2(g) + 2\,Ag^+(aq) \rightarrow 2\,H^+(aq) + 2\,Ag(s)$$

(A) There will be no change.
(B) It will be doubled.
(C) It will be halved.
(D) It will be decreased by an undetermined amount.

416. Which of the following statements is true concerning the reaction in an electrolytic cell?

(A) The reaction is spontaneous, and oxidation occurs at the cathode.
(B) The reaction is nonspontaneous, and oxidation occurs at the cathode.
(C) The reaction is spontaneous, and oxidation occurs at the anode.
(D) The reaction is nonspontaneous, and oxidation occurs at the anode.

417. Consider the following half-reaction potentials:

$IO_4^-(aq) + 2\,H^+(aq) + 2\,e^- \rightarrow IO_3^-(aq) + H_2O(l)$ $\quad E° = +1.65\ V$
$2\,SO_3^{2-}(aq) + 3\,H_2O(l) + 4\,e^-$ $\quad E° = -0.58\ V$
$\quad \rightarrow S_2O_3^{2-}(aq) + 6\,OH^-(aq)$
$Zn^{2+}(aq) + 2\,e^- \rightarrow Zn(s)$ $\quad E° = -0.76\ V$
$2\,Hg^{2+}(aq) + 2\,e^- \rightarrow Hg_2^{2+}(aq)$ $\quad E° = +0.91\ V$

Which of the following is the strongest reducing agent?

(A) $Zn(s)$
(B) $IO_4^-(aq)$
(C) $Hg_2^{2+}(aq)$
(D) $SO_3^{2-}(aq)$

418. An electrolytic cell contains molten $MgCl_2(l)$. Which of the following statements is true?

(A) Mg^{2+} is reduced at the anode, and Cl^- is oxidized at the cathode.
(B) Mg^{2+} is reduced at the cathode, and Cl^- is oxidized at the anode.
(C) Mg^{2+} is oxidized at the cathode, and Cl^- is reduced at the anode.
(D) Mg^{2+} is oxidized at the anode, and Cl^- is reduced at the cathode.

419. Consider the following reaction:

$$Ca(s) + Cl_2(g) \leftrightarrows CaCl_2$$

Which of the following is true concerning this reaction?
(A) Ca is an oxidizing agent.
(B) Cl_2 is a reducing agent.
(C) Cl_2 is an oxidizing agent.
(D) $CaCl_2$ is an oxidizing agent.

420. The following apparatus is used to electroplate silver onto a metal surface:

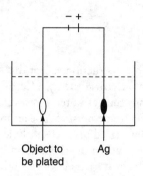

The object to be plated is the

(A) cathode, and it is negatively charged
(B) cathode, and it is positively charged
(C) anode, and it is negatively charged
(D) anode, and it is positively charged

421. The standard reduction potential for chlorine involves the following reaction:

$$Cl_2(g) + 2\ e^- \rightarrow 2\ Cl^-(aq) \qquad E° = +1.36\ V$$

How does this compare with the other halogens?
(A) $F_2 > Cl_2 = Br_2 > I_2$
(B) $F_2 = Cl_2 > Br_2 > I_2$
(C) $F_2 > Cl_2 > Br_2 > I_2$
(D) $F_2 < Cl_2 > Br_2 > I_2$

422. An electrolytic cell, with two inert electrodes, is used to electrolyze an aqueous potassium fluoride solution. The standard reduction potentials are:

$$F_2(g) + 2\ e^- \rightarrow 2\ F^-(aq) \qquad\qquad E^\circ = +1.36\ V$$
$$O_2(g) + 4\ H^+(aq) + 4\ e^- \rightarrow 4\ H_2O(l) \qquad E^\circ = +1.23\ V$$
$$2\ H_2O(l) + 2\ e^- \rightarrow H_2(g) + 2\ OH^-(aq) \qquad E^\circ = -0.83\ V$$
$$K^+(aq) + 1\ e^- \rightarrow K(s) \qquad\qquad E^\circ = -2.93\ V$$

What substance forms at the cathode?

(A) K(s)
(B) $H_2(g)$
(C) $O_2(g)$
(D) $F_2(g)$

423. Electrolytic cells are similar to voltaic cells in construction except that electrolytic cells require an external source of electrical energy. Electrolysis is capable of breaking apart many compounds such as water. However, to break down a nonelectrolyte, such as water, there must also be an inert electrolyte present. In an experiment, water is placed in an electrolytic cell, and a few drops of phosphoric acid are added. When the cell is connected to an external power source, gases begin to form at each electrode. What gas forms at the anode, and what gas forms at the cathode?

(A) $O^{2-}(g)$ at the anode and $H^+(g)$ at the cathode
(B) $H_2(g)$ at the anode and $O_2(g)$ at the cathode
(C) $O_2(g)$ at the anode and $PH_3(g)$ at the cathode
(D) $O_2(g)$ at the anode and $H_2(g)$ at the cathode

424. The standard cell potential for the following reaction is $E^\circ = +0.80$ V:

$$H_2(g) + 2\ Ag^+(aq) \rightarrow 2\ H^+(aq) + 2\ Ag(s)$$

What will happen to the cell potential if there is an increase in the pH?

(A) It will decrease.
(B) It will increase.
(C) It will remain the same.
(D) It is not possible to determine.

425. The cathode compartment of a voltaic cell involves the reduction of cupric ions. The reduction half-reaction for cupric ions is:

$$Cu^{2+} + 2\,e^- \to Cu(s) \qquad E° = 0.34\ V$$

How many faradays of electrons must pass through the cell to deposit 0.64 g of copper metal?

(A) 0.02
(B) 0.34
(C) 0.68
(D) 2.0

426. When the following redox equation is balanced, what is the coefficient for $Fe^{2+}(aq)$?

$$H^+(aq) + Cr_2O_7^{2-}(aq) + Fe^{2+}(aq) \to Cr^{3+}(aq) + Fe^{3+}(aq) + H_2O(l)$$

(A) 9
(B) 6
(C) 3
(D) 1

427. Which of the following compounds can serve only as an oxidizing agent?

(A) $KMnO_4$
(B) MnF_3
(C) MnO_2
(D) MnO

428. Which of the following oxidizing agents does NOT contain hexavalent chromium?

(A) K_2CrO_4
(B) CrO_3
(C) $K_2Cr_2O_7$
(D) $Cr_6(Si_6O_{18})$

429. The following reaction is pH dependent:

$$Cr_2O_7^{2-}(aq) + 2\,OH^-(aq) \leftrightarrows 2\,CrO_4^{2-}(aq) + H_2O$$

Which of the following statements is true?

(A) This is an oxidation.
(B) This is neither an oxidation nor a reduction.
(C) This is a reduction.
(D) This is a disproportionation.

430. The following reaction is spontaneous under standard conditions:

$$14\ H^+(aq) + 6\ Fe^{2+}(aq) + Cr_2O_7{}^{2-}(aq)$$
$$\rightarrow 6\ Fe^{3+}(aq) + 2\ Cr^{3+}(aq) + 7\ H_2O(l)$$

What can be said about the value of the Gibbs free energy change?

(A) $\Delta G = 0$
(B) $\Delta G > 0$
(C) $\Delta G < 0$
(D) ΔG is undefined.

431. The major source of magnesium metal is the electrolysis of molten magnesium chloride. In aqueous solution, the important half-reactions would be:

$$Mg^{2+}(aq) + 2\ e^- \rightarrow Mg \qquad E° = -2.37\ V$$
$$Cl_2(g) + 2\ e^- \rightarrow 2\ Cl^-(aq) \qquad E° = +1.36\ V$$

Assuming the reduction potential in molten magnesium chloride is the same as in aqueous magnesium chloride, what is the minimum voltage necessary for the electrolysis to occur?

(A) $-3.73\ V$
(B) $+3.73\ V$
(C) $-1.01\ V$
(D) $+1.01\ V$

432. How many moles of chlorine gas form during the formation of 8.8 g of strontium metal by the electrolysis of strontium chloride?

(A) 0.10 mol
(B) 8.8 mol
(C) 0.20 mol
(D) 0.40 mol

433. Identify the reducing agent in the following reaction:

$$8\ H^+(aq) + ClO_4{}^-(aq) + 7\ Cl^-(aq) \rightarrow 4\ Cl_2(g) + 4\ H_2O(l)$$

(A) $H^+(aq)$
(B) $ClO_4{}^-(aq)$
(C) $Cl^-(aq)$
(D) $Cl_2(g)$

434. A galvanic cell uses the following cell reaction:

$$3 \, In^{3+}(aq) + 2 \, Al(s) \rightarrow 3 \, In^{+}(aq) + 2 \, Al^{3+}(aq)$$

When the cell is operated, the measured standard potential was $+1.3$ V. The standard reduction potential for $Al^{3+}(aq)$ is -1.7 V. What is the standard reduction potential for the following half-reaction?

$$In^{3+}(aq) + 2 \, e^{-} \rightarrow In^{+}(aq)$$

(A) -0.4 V
(B) $+0.4$ V
(C) $+3.0$ V
(D) 0.0 V

435. The major method for producing sodium metal is through the electrolysis of molten sodium chloride. Copper, on the other hand, can be isolated by the electrolysis of aqueous copper chloride solutions. Why not produce sodium through the electrolysis of aqueous sodium chloride?

(A) The reaction is too slow to be viable.
(B) The process is too expensive.
(C) The other product, chlorine gas, reacts explosively with sodium metal.
(D) Sodium metal reacts with water.

436. The reference standard for half-reactions is the standard hydrogen electrode (SHE). The half-reaction for the SHE is:

$$2 \, H^{+}(aq) + 2 \, e^{-} \rightarrow H_2(aq)$$

Construction of a cell involves $1 \, M \, H^{+}(aq)$, 1 atm $H_2(g)$, and a platinum electrode. The reason why the platinum is necessary is that

(A) it is capable of undergoing reduction
(B) it is resistant to acid
(C) an electrical conductor is necessary
(D) it is easily oxidized

437. Which of the following reactions involves the oxidation of oxygen?

(A) $O_2{}^{2-}$ to O_2
(B) O_2 to H_2O
(C) OH^{-} to H_2O
(D) $O_2{}^{-}$ to $O_2{}^{2-}$

438. In general, any molecule containing 2 very electronegative atoms covalently bonded together will be a good oxidizing agent. Based on this generality, which of the following should be a good oxidizing agent?

(A) CF_4
(B) SO_2
(C) H_2O_2
(D) Na_2O

439. In the following redox reaction, what is the value of n?

$$14\ H^+(aq) + 6\ Fe^{2+}(aq) + Cr_2O_7^{2-}(aq)$$
$$\rightarrow 6\ Fe^{3+}(aq) + 2\ Cr^{3+}(aq) + 7\ H_2O(l)$$

(A) ⅙
(B) 6
(C) 14
(D) 1

440. Two of the half-reactions for the halogens are:

$$F_2(g) + 2\ e^- \rightarrow 2\ F^-(aq) \qquad E° = +2.87\ V$$
$$Cl_2(g) + 2\ e^- \rightarrow 2\ Cl^-(aq) \qquad E° = +1.36\ V$$

Based on these data, which of the following statements is true?

(A) Chlorine will oxidize fluorine.
(B) Both halogens are good reducing agents.
(C) Both halogens are good oxidizing agents.
(D) It is easier to oxidize fluorine than chlorine.

441. In a reduction half-reaction

(A) an element gains 1 or more electrons, and the oxidation state of the element increases
(B) an element loses 1 or more electrons, and the oxidation state of the element increases
(C) an element gains 1 or more electrons, and the oxidation state of the element decreases
(D) an element loses 1 or more electrons, and the oxidation state of the element decreases

442. A voltaic cell has the following two half-reactions:

$$(1)\ Sn^{2+}(aq) \rightarrow Sn^{4+}(aq) + 2\ e^-$$
$$(2)\ 2\ Fe^{3+}(aq) + 2\ e^- \rightarrow 2\ Fe^{2+}(aq)$$

Which of the following statements is correct?

(A) Reaction 2 is a reduction, and it occurs at the cathode.
(B) Reaction 1 is a reduction, and it occurs at the cathode.
(C) Reaction 2 is an oxidation, and it occurs at the anode.
(D) Reaction 1 is a reduction, and it occurs at the anode.

443. Consider the following half-reactions:

$$Sn^{4+}(aq) + 2\ e^- \rightarrow Sn^{2+}(aq) \qquad E° = +0.15\ V$$
$$Hg_2^{2+}(aq) + 2\ e^- \rightarrow 2\ Hg(l) \qquad E° = +0.79\ V$$

Which of the choices below is true about the following reaction?

$$2\ Hg(l) + Sn^{4+}(aq) \rightarrow Hg_2^{2+}(aq) + Sn^{2+}(aq)$$

(A) $E° = +0.64\ V$, and the reaction is spontaneous.
(B) $E° = -0.94\ V$, and the reaction is nonspontaneous.
(C) $E° = -0.64\ V$, and the reaction is nonspontaneous.
(D) $E° = +0.94\ V$, and the reaction is spontaneous.

444. Identify the oxidizing agent in the following reaction:

$$14\ H^+(aq) + 6\ Fe^{2+}(aq) + Cr_2O_7{}^{2-}(aq)$$
$$\rightarrow 6\ Fe^{3+}(aq) + 2\ Cr^{3+}(aq) + 7\ H_2O(l)$$

(A) $Fe^{2+}(aq)$
(B) $Cr_2O_7{}^{2-}(aq)$
(C) $Fe^{3+}(aq)$
(D) $Cr^{3+}(aq)$

445. What is the standard electrode potential for a galvanic cell using the following half-reactions?

$$IO_4^-(aq) + 2\,H^+(aq) + 2\,e^- \qquad E° = +1.65\text{ V}$$
$$\rightarrow IO_3^-(aq) + H_2O(l)$$
$$2\,SO_3^{2-}(aq) + 3\,H_2O(l) + 4\,e^- \qquad E° = -0.58\text{ V}$$
$$\rightarrow S_2O_3^{2-}(aq) + 6\,OH^-(aq)$$

(A) +1.07 V
(B) +1.65 V
(C) +2.23 V
(D) +3.88 V

446. If it were necessary to use 3 moles of HClO to oxidize 1 mole of Sn^{2+} in a sample to Sn^{4+}, how many moles of ClO_3^- would be necessary to perform the same oxidation?

$$HClO(aq) + H^+(aq) + 2\,e^- \rightarrow Cl^-(aq) + H_2O \qquad +1.49\text{ V}$$
$$ClO_3^-(aq) + 6\,H^+(aq) + 6\,e^- \rightarrow Cl^-(aq) + 3\,H_2O \qquad +1.45\text{ V}$$

(A) 1
(B) 3
(C) 2
(D) 6

447. The electrolysis of molten $MgCl_2$ produced 1.5 g Mg in 100 minutes. The reaction was:

$$MgCl_2(l) \rightarrow Mg(l) + Cl_2(g)$$

What was the approximate current used? (1 F = 96,500 C/mol)

(A) 0.20 A
(B) 0.50 A
(C) 1.0 A
(D) 2.0 A

448. It is possible to isolate calcium through the electrolysis of molten calcium chloride. In one experiment, an electrolytic cell electrolyzed a sample of molten calcium chloride. The cell operated at 5.0 V with a current of 2.0 amps. How many grams of calcium metal formed after 17 minutes? (1 F = 96,500 C/mol)

(A) 0.43 g
(B) 1.4 g
(C) 7.2 g
(D) 720 g

449. An analytical chemist constructs a galvanic cell. The cell uses the following two half-reactions:

$$HNO_2(aq) + H^+(aq) + e^- \rightarrow NO(g) + H_2O(l) \qquad E° = +1.00 \text{ V}$$
$$Ce^{4+}(aq) + e^- \rightarrow Ce^{3+}(aq) \qquad\qquad E° = +1.61 \text{ V}$$

She then measures the cell voltage under a variety of conditions. Which of the following conditions will yield the greatest voltage?

(A) Standard
(B) Low pH
(C) High pH
(D) Equilibrium

450. How many grams of mercury could be produced by the electrolysis of a $Hg(NO_3)_2$ solution using a current of 2.00 amps for 3.00 hours?

(A) 22.5 g
(B) 201 g
(C) 11.2 g
(D) 44.8 g

CHAPTER 12

Final Review

451. Water-soluble barium compounds are chronic toxins that are retained in the human body for years. Barium retention is due to barium ions mimicking what ion in humans?

(A) Ra^{2+}
(B) K^+
(C) Ca^{2+}
(D) H^+

452. What is the molar mass of $CuSO_4 \cdot 5H_2O$?

(A) 250 g/mol
(B) 180 g/mol
(C) 8,700 g/mol
(D) 160 g/mol

453. Which of the following pairs are compounds in which the halogen atom is in the same oxidation state?

(A) $(NH_4)_2SO_4$ and HNO_3
(B) Cl_2O and $HClO_2$
(C) Cl_2O_5 and $HClO_3$
(D) SO_3 and H_2SO_4

454. What is the concentration of nitrate ions in a solution prepared by adding 100 mL of 0.50 M calcium nitrate solution to 0.900 L of water?

(A) 0.080 M
(B) 0.056 M
(C) 0.050 M
(D) 0.10 M

455. Which of the following would have the highest heat of vaporization?

(A) $CH_3CH_2CH_2OH$
(B) $HOCH_2CH_2OH$
(C) $CH_3CH_2CH_2CH_3$
(D) $CH_3CH_2CH_2F$

456. The heat of vaporization does NOT depend upon which of the following?

(A) The external pressure
(B) The strength of the intermolecular forces
(C) The mass of the molecules
(D) The presence of hydrogen bonding

457. Which of the following has the highest melting point?

(A) H_2O
(B) NaCl
(C) CH_3OH
(D) $HOCH_2CH_2OH$

458. Which of the following contains the most moles of ions in solution?

(A) 1.0 mol $CHCl_3$
(B) 0.30 mol $CaCl_2$
(C) 0.45 mol KCl
(D) 0.25 mol $CeCl_3$

459. Dehydrated patients require intravenous saline solution. Why use saline solution instead of pure water?

(A) The osmotic pressure of water is too low.
(B) Dehydrated patients may have low salt levels.
(C) Saline solutions store better.
(D) Water becomes contaminated too easily.

460. Many substances dissolve in water because they can hydrogen bond to water. Which of the following CANNOT hydrogen bond to water?

(A) CH_3OH
(B) CH_3Cl
(C) $C_6H_{12}O_6$
(D) CH_3NH_2

461. Which of the following reactions will proceed at the greatest rate?
    (A) $A + B \rightarrow C$ E_a = 100 kJ
    (B) $D + E \rightarrow F + G$ E_a = 150 kJ
    (C) $H \rightarrow I + J$ E_a = 250 kJ
    (D) It is impossible to determine.

462. A kinetic study of the following reaction led to a proposed mechanism:

$$Cl_2(g) + H_2O(g) \leftrightarrows HCl(g) + HClO(g)$$

The proposed mechanism was:

$Cl_2 \leftrightarrows 2\ Cl\cdot$	fast
$Cl\cdot + H_2O \rightarrow HCl + \cdot OH$	slow
$Cl\cdot + \cdot OH \rightarrow HClO$	fast

The first step is promoted by light. Which of the following are intermediates?
    (A) $Cl\cdot$
    (B) $\cdot OH$
    (C) $Cl\cdot$ and $\cdot OH$
    (D) There is no intermediate.

463. One of the important antitumor agents is cisplatin. The formula for the compound is $Pt(NH_3)_2Cl_2$. The formation of this drug is an equilibrium process. Which of the following would increase the yield of this drug?
    (A) Adding $CHCl_3$ to increase the equilibrium chloride concentration
    (B) Adding hydroxide ion to reduce the equilibrium platinum concentration
    (C) Adding acid to react with the ammonia (a base)
    (D) Increasing the ammonia concentration of the equilibrium mixture

464. Calcium oxalate, CaC_2O_4, is a sparingly soluble compound. Which of the following would increase the solubility of this compound?
    (A) Increasing the pH
    (B) Lowering the pH
    (C) Lowering the temperature
    (D) Increasing the pressure

465. If the reaction quotient exceeds the K_{sp}

    (A) a precipitate will form
    (B) a precipitate will dissolve
    (C) the system will be in equilibrium
    (D) the results depend upon the temperature

466. The addition of a catalyst to a reaction

    (A) decreases the equilibrium constant
    (B) increases the equilibrium constant
    (C) does not change the equilibrium constant
    (D) It is impossible to determine the effect on the equilibrium constant.

467. Which of the following reactions is exothermic?

    (A) $A + B \rightarrow C$ $E_a = 100 \text{ kJ}$
    (B) $D + E \rightarrow F + G$ $E_a = -150 \text{ kJ}$
    (C) $H \rightarrow I + J$ $E_a = 250 \text{ kJ}$
    (D) It is impossible to determine.

468. The addition of a catalyst to a reaction

    (A) does not change the heat of reaction
    (B) increases the heat of reaction
    (C) decreases the heat of reaction
    (D) It is impossible to determine the effect on the heat of reaction.

469. The heat of reaction depends upon

    (A) the reactants and the transition state
    (B) the reactants, the transition state, and the products
    (C) the transition state and the products
    (D) the reactants and the products

470. The basic reaction of respiration is:

$$C_6H_{12}O_6(aq) + 6\,O_2(g) \rightarrow 6\,CO_2(g) + 6\,H_2O(l)$$

This is

    (A) a redox process
    (B) a reduction process
    (C) an oxidation process
    (D) a combination process

471. Normal chemical reactions continue until the limiting reagent completely reacts. However, batteries stop producing electricity while there is still limiting reagent remaining. Why is this true?

(A) As the concentrations decrease, the reaction rate decreases.
(B) The system has reached equilibrium.
(C) The reaction will not go to completion at room temperature.
(D) The presence of the products short-circuits the battery.

472. Cytochromes are a biologically important group of iron-containing molecules. They are important in redox processes. Part of the electron transfer process in the human body involves a transfer of an electron from cytochrome to cytochrome to cytochrome. The following are the standard reduction potentials for several cytochromes:

	$E°$
Cytochrome b-Fe^{3+} + e^- ⇆ Cytochrome b-Fe^{2+}	+0.08 V
Cytochrome c-Fe^{3+} + e^- ⇆ Cytochrome c-Fe^{2+}	+0.22 V
Cytochrome a-Fe^{3+} + e^- ⇆ Cytochrome a-Fe^{2+}	+0.29 V
Cytochrome a_3-Fe^{3+} + e^- ⇆ Cytochrome a_3-Fe^{2+}	+0.38 V

If each electron transfer is spontaneous, what is the sequence of electron transfer for the cytochromes in the table?

(A) a_3 to a to c to b
(B) b to c to a to a_3
(C) a_3 to c to b to a
(D) a_3 to a to b to c

473. The theoretical yield of NaCl for the following reaction depends upon which of the choices below?

$$2\ HCl(aq) + Na_2CO_3(aq) \rightarrow 2\ NaCl(aq) + CO_2(g) + H_2O(l)$$

(A) The amount of the limiting reagent
(B) The concentration of the excess reagent
(C) The amount of NaCl recovered
(D) The temperature

474. During nuclear fission, an atomic nucleus splits into two or more smaller nuclei and possibly other small particles. This process releases large amounts of energy. The fission is random, with not all nuclei splitting in the same way. It is possible to induce the fission of a uranium-235 atom with the absorption of a high-energy neutron. If one uranium-235 splits into a cesium-140 nucleus and releases three neutrons, what other nucleus will form?

(A) Rubidium-93
(B) Rubidium-92
(C) Strontion-90
(D) Strontium-92

475. Germanium is a semiconductor. The ground-state electron configuration of a germanium atom is

(A) $[Ar]4s^23d^{10}4p^2$
(B) $[Ar]4s^23d^94p^3$
(C) $[Ar]4s^23d^64p^6$
(D) $[Ar]3d^{10}4p^4$

476. Mirror nuclei are nuclei in which one has the same number of neutrons as the other has protons and the same number of protons as the other has neutrons. Which of the following pairs are mirror nuclei?

(A) ^{27}Al and ^{27}Si
(B) ^{23}Na and ^{24}Mg
(C) ^{48}Ti and ^{48}Ca
(D) ^{32}P and ^{36}Cl

477. Which of the following is an excited-state electron configuration of sodium?

(A) $1s^22s^22p^63p^1$
(B) $1s^22s^22p^63s^1$
(C) $1s^22s^22p^6$
(D) $1s^22s^22p^53s^1$

478. Ionization energy increases from left to right on the periodic table. Which other trends also occur in this direction?

(A) Increasing electronegativity and decreasing atomic radii
(B) Decreasing electronegativity and decreasing atomic radii
(C) Increasing electronegativity and increasing atomic radii
(D) Decreasing electronegativity and increasing atomic radii

479. Which of the following is NOT a polar molecule?

(A) HCl
(B) CH_3Cl
(C) H_2S
(D) CF_4

480. The BEST description of the orbital geometry of an sp hybridized carbon atom is

(A) tetrahedral
(B) linear
(C) trigonal planar
(D) trigonal pyramid

481. Which of the following elements is the LEAST reactive?

(A) Ca
(B) Be
(C) Sr
(D) Ra

482. Which of the following ionic compounds will have the lowest melting point?

(A) RbI
(B) AlN
(C) CaO
(D) LiF

483. A gas mixture with a total pressure of 1,000 torr at 25°C contains 3 moles of O_2, 1 mole of H_2, and 4 moles of N_2. What is the partial pressure of H_2?

(A) 500 torr
(B) 1,000 torr
(C) 125 torr
(D) It is impossible to determine without knowing the volume of the system.

484. The ratio of the rate of effusion of methane gas is what multiple of the rate of effusion of oxygen gas?

(A) 1.4
(B) 0.25
(C) 1
(D) $32/16$

485. A sample of helium gas has a volume of 500 cm³ at 227°C. What is the volume at 727°C? Assume that the pressure remains constant and that helium is nearly ideal under these conditions.

(A) 1,000 cm³
(B) 500 cm³
(C) 250 cm³
(D) 750 cm³

486. A "perfect" gas is an ideal gas; however, many substances approach ideal behavior under certain circumstances. Water vapor approaches ideal behavior when

(A) the pressure is low and the temperature is high
(B) the pressure is high and the temperature is high
(C) the pressure is low and the temperature is low
(D) the pressure is high and the temperature is low

487. $Ce(NO_3)_3$ is soluble in water. Approximately how many moles of $Ce(NO_3)_3$ are necessary to produce 1.0 mole of ions in solution?

(A) 0.08
(B) 1.0
(C) 0.25
(D) 4.0

488. The following reaction obeys the rate law Rate = k $[Br_2][NO]^2$:

$$Br_2(g) + 2\ NO(g) \rightarrow 2\ NOBr(g)$$

Which of the following might be the mechanism for this reaction?

(A) $Br_2(g) + NO(g) \rightarrow NOBr_2(g)$ fast
    $NOBr_2(g) + NO(g) \rightarrow 2\ NOBr(g)$ slow
(B) $Br_2(g) + 2\ NO(g) \rightarrow 2\ NOBr(g)$
(C) $Br_2(g) + NO(g) \rightarrow NOBr_2(g)$ slow
    $NOBr_2(g) + NO(g) \rightarrow 2\ NOBr(g)$ fast
(D) $Br_2(g) \rightarrow 2\ Br(g)$ slow
    $2\ [Br(g) + NO(g) \rightarrow NOBr(g)]$ fast

489. The following equilibrium occurs at a certain temperature:

$$2\ NO_2(g) \leftrightarrows 2\ NO(g) + O_2(g)$$

After equilibrium is established, a small quantity of NO_2 is added, causing the

(A) reaction to shift to the right
(B) reaction to shift to the left
(C) equilibrium constant to increase
(D) equilibrium constant to decrease

490. Which of the following acids would be the BEST choice for preparing a buffer solution of 8.6 at 25°C?

(A) $HBrO(aq)$ ($pK_a = 8.60$)
(B) $HClO(aq)$ ($pK_a = 7.30$)
(C) $HF(aq)$ ($pK_a = 3.17$)
(D) $HIO(aq)$ ($pK_a = 10.64$)

491. The conjugate bases of $H_2PO_4{}^-$, CH_3OH, and NH_3, respectively, are

(A) $PO_4{}^{3-}$, CH_3O^-, and $NH_2{}^-$
(B) $HPO_4{}^{2-}$, CH_3O^-, and $NH_2{}^{2-}$
(C) $HPO_4{}^{2-}$, CH_3O^-, and $NH_2{}^-$
(D) H_3PO_4, $CH_3OH_2{}^+$, and $NH_4{}^+$

492. The K_a of HOCN is 3.5×10^{-4}. Which of the following calculations will give the pK_b of the cyanate ion?

(A) $-14 + \log(3.5 \times 10^{-4})$
(B) $14 - \log(3.5 \times 10^{-4})$
(C) $14 + \log(3.5 \times 10^{-4})$
(D) $-14 - \log(3.5 \times 10^{-4})$

493. The K_a for HNO_2 is 5.1×10^{-4}. What is the approximate pK_a?

(A) 3.3
(B) 4.1
(C) 5.0
(D) 1.0

494. The following reaction will form phosgene from carbon monoxide and chlorine:

$$CO(g) + Cl_2(g) \rightarrow COCl_2(g)$$

Consider the following thermodynamic data:

Compound	ΔH_f° (kJ/mol)	ΔG_f° (kJ/mol)	S° (J/mol·K)
$CO(g)$	−110	−137	198
$Cl_2(g)$	0	0	223
$COCl_2(g)$	−223	−210	290

The entropy change for the formation of 1 mole of phosgene is

(A) −131 J/K
(B) −290 J/K
(C) +290 J/K
(D) −103 J/K

495. Consider the following reaction:

$$2\,Al(s) + 6\,H^+(aq) \rightarrow 2\,Al^{3+}(aq) + 3\,H_2(g)$$

What will happen as the reaction proceeds?

(A) The pH will increase.
(B) The pH will decrease.
(C) The [Al] will increase.
(D) The pH will remain constant.

496. A container half-filled with ethanol, C_2H_5OH, is connected to a vacuum pump, and the air is removed from the container. How does this affect the ethanol?

(A) More ethanol molecules dissociate.
(B) The vapor pressure of ethanol increases.
(C) The boiling point of ethanol decreases.
(D) The ethanol decomposes.

497. Chlorine dioxide is an industrial bleach and a free radical. The Lewis structure for chlorine dioxide is

(A) 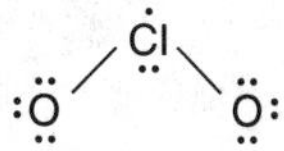 $\ddot{:}\ddot{O}\diagup\overset{\ddot{C}l}{}\diagdown\ddot{O}\ddot{:}$

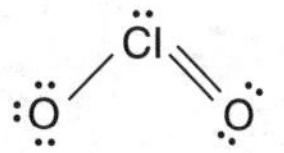

(B) $\ddot{:}\ddot{O}\diagup\overset{\ddot{C}l}{}\diagup\!\!\diagdown\ddot{O}$

(C) $\ddot{O}\diagup\overset{Cl}{}\diagup\!\!\diagdown\ddot{O}$

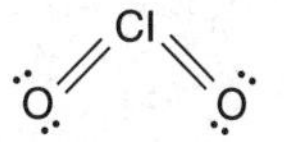

(D) 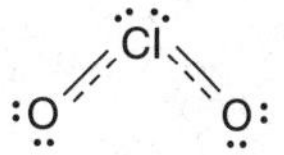 $\ddot{:}O\diagup\overset{\ddot{C}l}{}\diagdown O\ddot{:}$

498. In the presence of a catalyst, the following reaction occurs:

$$CO_2(g) + 2\,H_2(g) \rightarrow CH_2O(g) + H_2O(g)$$

What are the geometries on the carbon compounds?

(A) CO_2 is bent, and CH_2O is trigonal pyramidal.
(B) CO_2 is linear, and CH_2O is trigonal pyramidal.
(C) CO_2 is bent, and CH_2O is trigonal planar.
(D) CO_2 is linear, and CH_2O is trigonal planar.

499. How many half-lives have passed if only 6.25% of a certain radioactive isotope remains?

(A) 6
(B) 2
(C) 3
(D) 4

500. Phosphoric acid (H_3PO_4) is a triprotic weak acid, which leads to its having three K_a values. The approximate K_a values for phosphoric acid are $K_{a1} = 8 \times 10^{-3}$, $K_{a2} = 6 \times 10^{-8}$, and $K_{a3} = 4 \times 10^{-13}$. The approximate value of pK_{a1} is

(A) 7.0
(B) 7.2
(C) 12.5
(D) 2.1

ANSWERS

Chapter 1: The Basics

1. (D) A mole of H_3PO_4 would contain 3 moles of hydrogen; 5 moles of phosphoric acid would contain 15 moles of hydrogen. Hydrogen has a molar mass of about 1 g/mol; 15 moles would have a mass of 15 g.

2. (B) 1 mole of sulfuric acid has an approximate mass of 98 g/mol. In 1 mole of sulfuric acid (approximately 100 g), you will have 2 moles of H (2 g, or 2%), 1 mole of S (32 g, or 32%), and 4 moles of O (4 mol $\times$ 16 g/mol = 64 g, or 64%). Answer B is the closest.

3. (C) From the balanced equation, you can see that decomposition of 2 moles of copper(II) nitrate will yield 4 moles of NO_2, so (4 mol $\times$ 46 g/mol = 184 g) about 180 g will be formed.

4. (A) The sum of all oxidation numbers in the compound must equal zero, with common oxidation states for potassium and oxygen to be +1 and −2, respectively. Therefore, (2 $\times$ +1) + Fe + (4 $\times$ −2) = 0; the oxidation state of iron is +6.

5. (A) Sulfur has an oxidation state of −2, so since there are 2 copper atoms in answer A, each must be in the +1 oxidation state. Good test-taking strategies would indicate that since there is no answer "A and B," etc., then answer A would be the only answer having the +1 species, and there is no reason to calculate the others.

6. (B) The formula mass of pyruvic acid is 88.1 g/mol. In 1 mole of pyruvic acid, there are 3 moles of oxygen, or 48 g of oxygen. The percentage oxygen would be (48 g / 88.1 g) $\times$ 100% = 55%.

7. (A) The reaction is not a neutralization (no traditional acid or base involved with the pH approaching 7), nor is it a single replacement, since that would involve solid iron replacing the hydrogen. The oxidation states of the elements involved in the reaction have not changed, so it is not an oxidation-reduction; therefore, it is a Lewis acid-base reaction.

8. (A) At high temperatures, both calcium carbonate and sodium carbonate decompose, leaving the respective oxides and releasing carbon dioxide. Na_2O is not a gas, and both oxygen and ozone would be very reactive at these temperatures, eliminating those choices.

9. (C) The formula of ammonium phosphate is $(NH_4)_3PO_4$. For every mole of ammonium phosphate, there are 3 moles of ammonium ions. The solution is 0.3 M in ammonium phosphate. The concentration of ammonium ions would be three times that concentration, or 0.9 M.

10. (D) The net ionic reaction is $OH^- + H^+ \rightarrow H_2O$. The concentration of the hydroxide ion in the sodium hydroxide solution is 0.40 M, while the concentration of the hydronium ion in the hydrochloric acid solution is 0.20 M. The OH^- and the H^+ are reacting in a 1:1 ratio, but the sodium hydroxide solution is twice as concentrated as the hydronium ion solution; therefore, it will require half the volume of the base solution to supply the same number of hydroxide ions as hydronium ions. There is 50.0 mL of acid (0.0500 L), so it will require half that amount, 25 mL, of base.

11. (A) This is a good problem for using estimations. Eleven moles of oxygen would have a mass of 176 g [$(10 \times 16) + 16$], while the 12 moles of carbon and 22 moles of hydrogen would have a mass of 166 g [$(12 \times 12) + 22$]. Those are close enough to being equal to determine that the percent oxygen is slightly above 50%. The actual percentage is 176 g / 342.3 g = 51.4% oxygen.

12. (A) Molarity is moles of solute per liter of solution. A density of 1 g/mL would also be 1,000 g/L. The molar mass of water is 18 g/mol. The number of moles of water in 1,000 g of water is (1,000 g)/(18 g/mol) = 55 mol. 55 mol/1 L = 55 M.

13. (A) Since this is a neutral compound, the sum of the oxidation states must be zero. Hydrogen is +1 and oxygen is −2, so for HCHO it will be (+1) + C + (+1) + (−2) = 0. Therefore, carbon is 0.

14. (D) In the periodate ion, the sum of all oxidation states must be −1 (the charge on the polyatomic ion). Oxygen is −2, so for IO_4^-, it will be (I) + (4 × −2) = −1; I is +7.

15. (A) Reason through using approximations without working through all the math. The molar masses of Re_2O_7 and H_2 are 484 g/mol and 2 g/mol, respectively. Therefore, 1.0 kg for each amounts to a little over 2 moles of Re_2O_7 and 500 moles of H_2. After comparing the amounts given, it is obvious that $Re_2O_7(s)$ is the limiting agent. Slightly over 2 moles of Re_2O_7 should produce slightly over 4 moles of Re. The amount of Re formed should be a little over 4 × 186 g = 744 g (= 0.744 kg).

16. (C) The sulfuric acid solution, since it is diprotic, can provide (0.100 L × 6 mol/L H^+) 0.6 mol H^+, and the potassium hydroxide solution can provide (0.200 L × 3 mol/L OH^-) 0.6 mol OH^-. There is complete neutralization so that remaining in solution is 0.3 mol SO_4^{2+} + 0.6 mol K^+ = 0.3 mol K_2SO_4 contained in 300 mL. 0.3 mol K_2SO_4/0.300 L = 1 M.

17. **(B)** Chlorine is in the $+7$ oxidation state in Cl_2O_7 $[(2 \times Cl) + (7 \times -2) = 0]$. In HCl, Cl is in the -1; in $HClO_3$, Cl is in the $+5$; and in Cl_2, Cl is in the 0 state. In $HClO_4$, Cl is in the $[+1 + Cl + (4 \times -2) = 0]$ $+7$ oxidation state.

18. **(A)** Oxygen is in the -2 oxidation state, and there are 4 of them, giving a -8. The 3 irons must share a total oxidation state of $+8$, making each iron a $+8/3$.

19. **(C)** The empirical formula has the elements in the compound in the lowest whole-number ratio. With a molecular formula of $C_6H_{12}O_6$, dividing each subscript by 6 leaves CH_2O.

20. **(D)**

$$\frac{0.20 \text{ mol HCl}}{L} \times \frac{0.05000 \text{ L}}{1} \times \frac{1 \text{ mol Ba(OH)}_2}{2 \text{ mol HCl}} \times \frac{1 \text{ L}}{0.20 \text{ mol Ba(O)}_2}$$
$$= 0.025 \text{ L (25 mL)}$$

21. **(A)** There are $(0.250 \text{ L} \times 0.10 \text{ mol/L})$ 0.0250 moles of OH^- added to $(0.250$ L $\times$ 0.20 mol/L$)$ 0.050 moles of H^+. The acid-base reaction occurs in a 1:1 ratio, leaving 0.025 moles H^+ contained in 500.0 mL. The concentration of H^+ would be 0.025 mol/0.500L $= 0.050$ M.

22. **(B)** The molar mass of C_2H_5OH is 46 g/mol. There are $6.02 \times 1,023$ molecules per mole. $3.01 \times 1,023$ molecules therefore would be 0.5 moles. 46 g/mol $\times$ 0.5 mol $= 23$ g.

23. **(C)** Molality is moles solute per kilogram of solvent. The moles of the solute would be 16 g / 32 g/mol $= 0.5$ mol. Molality would be 0.5 mol / 0.500 kg $= 1.0$ m.

24. **(A)** According to the reaction stoichiometry, 3 moles of H_2SO_4, 2 moles of $KMnO_4$, and 5 moles of $H_2C_2O_4$ will react, leaving 1 mole of sulfuric acid. The molar mass of sulfuric acid is 98 g/mol, so 1 mole will be about 100 g.

25. **(B)** The sum of the oxidation states of all the elements in a compound will be zero. Hydrogen is $+1$ and oxygen is -2, so carbon must be 0.

26. **(C)** A mole of P_4 would have a mass of approximately 120 g $(4 \times 30 \text{ g/mol})$. Therefore, 1.5 moles would have a mass of 180 g.

27. **(A)** Assume 100 g of the compound, and convert to approximate number of moles using rounded-off masses and molar masses. Ag: 80 g / 100 g/mol $= 0.8$ mol Ag; P: 8 g / 32 g/mol $= 0.25$ mol P; O: 16 g / 16 g/mol $= 1$ mol O. Divide each number of moles by the smallest (0.25), giving a Ag/P/O ratio of 3.2/1/4, which is closest to answer A.

28. (A) In both compounds, the sum of the oxidation numbers must equal zero; the common oxidation numbers of hydrogen and chlorine are $+1$ and -2, respectively. Therefore, in $HClO_2$, you would have $(+1) + Cl + (2 \times -2) = 0$; the oxidation number of Cl would be $+3$. In $HClO_4$, you would have $(+1) + Cl + (4 \times -2) = 0$; the oxidation number of Cl would be $+7$.

29. (D) For every 2 moles of Ag_2O that decompose, you get 1 mole of oxygen gas; 3.0 moles will yield 1.5 moles of oxygen gas.

30. (B) The formula mass of ethyl alcohol is 46.1 g/mol. In 1 mole of ethyl alcohol (46.1 g), there are 2 moles of carbon (24.01 g). The percentage carbon would then be (24.01 g / 46.1 g) $\times$ 100% = 52% carbon. This eliminates answers C and D. The percent oxygen would be (16.0 g / 46.1 g) $\times$ 100% = 35%, eliminating answer A.

31. (D) AgI is an ionic compound, composed of Ag^+ and I^-. There has been no change in the iodide ion, but silver has gone from Ag^+ to solid silver by gaining an electron. The gain of electron(s) is reduction. You have no electron pair donors or receivers, so that eliminates the Lewis acid-base reaction, and nothing has become an ion that wasn't already an ion.

32. (A) Formal charge of an atom = # of valence electrons − (# of unshared valence electrons + ½ # of shared valence electrons). The cyanide ion has the Lewis structure of $:C:::N:^-$. Therefore, the formal charge on carbon is $4 - (2 + 3) = 4 - 5 = -1$; the formal charge on nitrogen is $5 - (2 + 3) = 5 - 5 = 0$.

33. (C) The reaction releases hydrogen ions; therefore, the pH should be lower than before the reaction took place. The pH of the final solution was 5, so the pH was initially above 5.

34. (C) Molarity is moles of solute per liter of solution. 100.0 cm^3 is 100.0 mL = 0.1000 L. Therefore, the molarity would be 0.2 mol solute / 0.1000 L = 2.0 M.

35. (D) Molality is moles of solute per kilogram of solvent. There is 1,000 g of the water solvent; that is 1 kg solvent. 5 mol ethanol / 1 kg solvent is a 5 m solution.

36. (A) Since there are 2 moles of sodium per every mole of sodium carbonate, the original sodium ion concentration in the first solution would be 0.20 M. The actual number of moles of sodium would be 0.0500 L $\times$ 0.20 mol/L = 0.01 mol. The moles of sodium ion in the sodium chloride solution would be 0.0500 L $\times$ 0.10 mol/L = 0.005 mol. The two solutions are mixed. There are now 0.015 moles of Na^+ contained in 100 mL of solution. The molar concentration would therefore be 0.015 mol / 0.100 L = 0.15 M.

37. (B) The formula of water is H_2O with a molar mass of 18 g/mol. An Avogadro's number of particles is 1 mole of particles, and in the case of water, it would have a mass of 18 g.

38. (C) Since amounts of both reactants were given, you must first determine which is the limiting reactant. This can be accomplished by dividing the moles of reactants by their respective coefficient in the balanced chemical equation. For the HCl, it would be 2.0 mol/2 = 1, and for the FeS, it would be 0.5 mol/1 = 0.5. Therefore, the FeS is the limiting reactant, since it has the smaller mole-to-coefficient ratio. Now basing the stoichiometry on the FeS, you see that for every mole of FeS that reacts, 1 mole of H_2S is formed. 0.5 moles of FeS reacts, forming 0.5 moles of H_2S.

Chapter 2: Atomic Structure

39. (C) In electron capture, an electron combines with a proton to form a neutron. Therefore, the atomic number (the number of protons) decreases by 1, from 20 to 19, from Ca to K.

40. (C) Answers A and B refer to the actual mass of an atom, while answer D refers to the mass in grams of a mole of a substance. The mass defect refers to the difference between the mass of an atom (isotope) and the sum of the masses of its protons, neutrons, and electrons.

41. (B) The electron configuration indicates that there are 12 electrons, and the problem states that it is an element involved, not an ion, so that there are also 12 protons. The element with an atomic number of 12 is magnesium.

42. (A) The ns^2np^6 configuration would indicate the presence of p electrons, which is true of all the noble gases except He.

43. (C) Gold (Au) has 79 protons. In alpha decay, the element loses 2 protons and 2 neutrons. Therefore, the number of protons (atomic number) of the new isotope will be 77, the element Ir.

44. (C) The electron configuration indicates that the element has 36 (Kr kernel) + 9 = 45 electrons. The element with atomic number 45 is Rh.

45. (D) Radioactive decay is first-order kinetics, where $t_{1/2} = 0.693/k$. If you make a fraction of the half-life expression Na-30/Cu-57, you would have Na-30/Cu-57 = 50 ms/200 ms = $0.693/k_{Na}/0.693/k_{Cu}$. The 0.693 cancel, leaving you with a ratio of k_{Na}/k_{Cu} of 4:1.

46. (B) In going from Br-36 to Kr-36, the mass number remains constant, but the atomic number increases by 1. This means that a neutron was converted into a proton, releasing an electron (β^- decay).

47. (B) In electron capture, a proton combines with an electron to form a neutron. The atomic mass remains constant (eliminating answers C and D), and the atomic number decreases by 1. Pd is atomic number 46; therefore, the isotope with atomic number 45 is produced, Rh.

48. (C) 40 days = 5 half-lives (40 days/8 days per half-life). The half-life decay sequence goes 50%, 25%, 12.5%, 6.25%, and 3.13% at 5 half-lives.

49. (D) In the alpha decay process, two protons and two neutrons are emitted. Therefore, the atomic number decreases by two. In order for the same element to be produced at the end, two protons must be created. In beta decay, a neutron is converted into a proton and an electron (beta particle). It will take two beta decays to produce the two protons needed.

50. (B) An electric field will affect any species that contains a charge. All of the answers contain charged species except answer B, gamma rays.

51. (B) The sample would decay like this: 50% → 25% → 12.5% → 6.25% → 3.125%, and so on. The closest to 5% is the 6.25%, 4 half-lives. It would actually take a little longer than 4 half-lives to reach the 5%. Four half-lives are 32 minutes (4 × 8 min), so the 35 minutes is the closest.

52. (A) In all these compounds and in elemental nitrogen, the sum of the oxidation numbers must equal zero, since the compounds are neutral. Hydrogen is in the +1 state and oxygen is −2. Therefore, in answer A: $(+1) + N + (3 \times -2) = 0$, so N is +5; in answer B: $N + (3 \times +1) = 0$, so $N = -3$; in answer C, an element, nitrogen, is in the zero state; and in answer D: $(2 \times +1) + 2N + (2 \times -2) = 0$, so $N = +1$. Answer A is the highest.

53. (C) Silver metal would have 11 electrons beyond the Kr kernel. The cation would have only 10, all in the 4d. Answer B represents the ground state of the silver ion. However, in the excited state, an electron would be promoted to a higher energy level, such as the 5s. Answer A is the electron configuration of silver metal, not the silver cation. Answer D is the ground state of Ag^{2+}. Answer C has the required 10 electrons, with 1 in the 5s.

54. (D) In beta emission, a neutron is converted into a proton; the atomic number increases by 1, but the mass number remains the same. In the beta decay of tritium, increasing the atomic number by 1 would produce He, and it should be ^3He because the mass number stays the same.

55. (A) Two beta decays would increase the atomic number by 2, but losing the beta particle would decrease it by 2—there would be no net change in atomic number. However, the mass number should decrease by 4, from 228 to 224.

56. (D) In the n = 4 shell, you will have the 4s (2 electrons), the 4p (6 electrons), the 4d (10 electrons), and the 4f (14 electrons). That gives 32 electrons.

57. (C) For gold, n = 6, l values will range from 5 to 0. The m_l values will be the ones that contain a quantum number of −3. The first one that is possible is the situation in which l = 3, an f orbital, and specifically, the 4f orbital.

58. (B) Starting with 1 g, the decay sequence would be 1 g → 0.5 g → 0.25 g, and so on. The 0.23 g is closest to two half-lives, or 2 × 5,700 years = 11,400 years.

59. (D) Alpha particles are composed of two protons and two neutrons, a He-4 nucleus. If they gained a couple of electrons, helium gas forms.

60. (A) *Degenerate* is a term that refers to orbitals that have the same energy.

61. (A) In beta emission, a neutron is converted into a proton and a beta particle that is expelled from the nucleus. Therefore, the atomic number of the product is one unit higher than that of the particle that underwent beta emission. Iodine has an atomic number of 53. The product would have an atomic number of 54, which is Xe.

62. (A) The Bohr model worked well for atoms or ions that have a single electron. Those that have a single electron are Bohr atoms. He^+ has lost one of its two electrons, leaving one. H^- has two electrons, and H^+ has none. Li^+ forms when lithium loses its valence electrons, leaving two.

63. (D) ^{15}N has a "magic number" of neutrons (8). None of the others has a magic number of either protons or neutrons.

64. (C) Fe (Z = 26) does not have a "magic number" of protons and is therefore eliminated. The remaining answers have magic numbers of protons. Sn-118 would have (118 − 50) 68 neutrons and is eliminated. O-17 has (17 − 8) 9 neutrons and is eliminated. Pb-208 would have (208 − 82) 126 neutrons, which is a magic number, and so answer C is correct.

65. (C) Xe-128 has (128 − 54) 74 neutrons, and Ba-130 has (130 − 56) 74 neutrons; they are isodiapheres, while none of the other pairs have the same number of neutrons.

66. (D) Beta particles are very penetrating and can do significant damage. Electron capture gives off x-rays, which can be very damaging. Alpha particles are not very penetrating, so answers A and D are possible. But answer D has a very long half-life, so only a small amount of radiation is released per unit of time.

67. (C) The amount of radioactivity would reduce at the following rate: 1 $t_{\frac{1}{2}}$ (3.8d) = 5 × 10⁻⁶; 2 $t_{\frac{1}{2}}$ (7.6d) = 2.5 × 10⁻⁶; 3 $t_{\frac{1}{2}}$ (11.4d) = 1.25 × 10⁻⁶; 4 $t_{\frac{1}{2}}$ (15.2d) = 6.25 × 10⁻⁷. The radiation level will fall below 1 × 10⁻⁶ a little after 11.4 days, so 12 days is the most reasonable.

68. (D) In beta emission, a neutron is converted into a proton with emission of a beta particle, thus reducing the number of protons and increasing the number of protons. This decreases a high neutron/proton ratio.

69. (A) As the electrons drop from a higher energy state to a lower one (3p → 1s), energy is released (this eliminates answers C and D) and shows up as a bright band in the atomic spectrum.

70. (A) An alpha particle consists of 2 protons and 2 neutrons (4 units of mass). Two alpha particles would have twice the number of units of mass, or a total of 8 units of mass.

71. (A) An s subshell contains 1 orbital; a p subshell contains 3 orbitals; a d subshell contains 5, and so on—a progression of adding 2 to the previous number.

72. (D) Iron would have 26 electrons, while the Fe^{3+} cation would have 23 electrons. All answers use the [Ar] core (18 electrons), so the ion would have an additional 5 electrons (18 + 5 = 23). That would be answers B and D. However, the 4s electrons leave before the 3d electrons, so the results will have no 4s electrons.

73. (A) In one half-life, you would have 50% remaining, but 8 years is a little over half of a half-life, so the correct answer must be greater than 50%. This eliminates answers B and D. Radioisotopes decay as the log function, not a linear function, so you should have less than 75% remaining in 8 years. Answer A is the only one that fits this criterion.

74. (C) Paramagnetism results from having unpaired electrons. Answers A and D have an unpaired s-electron, while answer B has an unpaired d-electron. In the Cu^+, there are only paired electrons.

75. (A) Silicon has 14 electrons. (This eliminates all the answers except answer A, because the others show 15 electrons.) The electron configuration of silicon in its ground state would be $1s^2 2s^2 2p^6 3s^2 3p^2$. It would be one of the 3p electrons that would be promoted to a higher energy state, such as the 4s.

76. (D) Thorium is atomic number 90 and lead is 82. For every alpha particle that is emitted, the atomic number decreases by 2, and for every beta particle emitted, it increases by 1. The 7 alpha decays will take the atomic number from 90 to 76, and the 6 beta decays will increase it from 76 to 82.

77. (B) The amount of energy produced can be calculated by Einstein's equation, $E = mc^2$. The speed of light, c, is 3.0×10^8 m/s. The units of $J = kg \cdot m/s$, so the mass must be expressed in kg, 2.0×10^{-31} kg. Therefore, $E = (2.0 \times 10^{-31}) (3.0 \times 10^8)^2 = 1.8 \times 10^{-14}$ J.

78. (C) As the two nuclei fuse, energy is released. That energy comes from a small amount of mass being converted into energy. Therefore, the nucleus formed would have a mass slightly smaller than the sum of the other two nuclei.

79. (D) Sodium has a single valence electron in the 3s orbital. (This eliminates answer C, where the energy level is shown as 2.) The valence electron is in an s orbital ($l = 0$), so answer A is eliminated. The only value of m_l with $l = 0$ would be 0, eliminating answer B.

80. (D) No two electrons may have the exact same set of four quantum numbers.

81. (B) Sp^2 hybridized carbons are those involved in double bonds. In the fused ring system, there are 5 carbons that are part of a double bond.

82. (C) Fe_2O_3 is commonly called rust and is certainly colored. It is the only compound with a partially filled d subshell.

83. (D) Taking a diuretic would not decrease the iodine-131, since it tends to concentrate in the thyroid and would eliminate both isotopes. Boiling will not affect the rate of decay of any radioisotope. Adding silver nitrate would precipitate both iodine isotopes. The only choice that would work is to increase the intake of nonradioactive iodine, thus diluting the radioactive form. That would increase the possibility of nonradioactive iodine being absorbed.

Chapter 3: Bonding and the Periodic Table

84. (C) Potassium and arsenic are in the same period. The atomic radius of an atom decreases as you move from left to right in a period because the increasing effective nuclear charge pulls the valence electrons closer to the nucleus. Therefore, potassium would be larger than arsenic due to the increasing effective nuclear charge.

85. (A) Magnesium chloride is an ionic compound (metal + nonmetal), and therefore no electrons are shared—they have been transferred.

86. (A) The third ionization energy of calcium will be much higher than that of scandium, since in calcium, you have removed the 4s electrons and are now trying to remove a 3p electron (core electron) that is much closer to the nucleus. With scandium, the three electrons are in the same energy level and do not require as much energy.

87. (A) There should be 12 electrons (6 from each oxygen) in the Lewis structure of diatomic oxygen. Only structure A has 12 electrons being used, with each oxygen having a full octet. (Counting electrons and bonds is important while taking the MCAT.)

88. (B) The emission of light is due to the electron(s) dropping from a higher energy state to a lower one, since the electrons must release energy. To reach an excited state, energy must be absorbed; that energy is released when those electrons drop back down to their ground state. Each transition would have a certain energy and a certain wavelength associated with it.

89. (A) The term *normal* in this case refers to standard conditions of pressure, which is 1 atmosphere, or 760 torr.

90. (A) The nitride ion is N^{3-}, formed by the gain of three electrons. Using the approximate 20% larger rule, the first electron would increase the radius of 75 pm to 90 pm. The second electron would increase the radius from 90 pm to 108 pm. The addition of the third electron would take it from 108 pm to 130 pm. Answer A is the closest.

91. (C) Since the speed is inversely proportional to the ionic radius, the slowest (smallest) rate of diffusion would be by the ion with the largest ionic radius. This would be the barium ion, since it is in period 6.

92. (B) Ionization energies, the energy needed to remove a valence electron from a gaseous atom, are related to the size of the atom. The farther the electron is from the nucleus, the lower the ionization energy. Moving from left to right in a period, the ionization energies increase, since the increasing effective nuclear charge is pulling the valence electrons closer to the nucleus. This eliminates answers A and D. As you proceed down a group, the ionization energy decreases, since the valence electrons are farther from the nucleus and it requires less energy to remove them. This eliminates answer C.

93. (D) During the first electron affinity, a negative electron is being added to a neutral atom. However, during the second electron affinity, a negative electron is being added to an anion that already has a negative charge. The repulsion of the like charges must be overcome to add the second electron, and thus the second electron affinity value is much greater than the first.

94. (B) A lower ionization energy is related to the fact that the electron being lost is farther from the nucleus. Sodium and chlorine are in the same period. Because of the increasing effective nuclear charge as you go from left to right on the periodic table, the atoms get smaller and the ionization energy increases, since the valence electrons are closer to the nucleus. The chlorine atom is smaller than the sodium atom. The valence electron is farther from the nucleus than in chlorine, and thus, sodium has a smaller ionization energy.

95. (C) This binding does not involve a redox reaction, eliminating answers A and D. The oxygen molecule has several pairs of valence electron that can be donated to the iron. This makes the oxygen a Lewis base, donating a pair of electrons.

96. (D) In the bonding of oxygen to the iron in hemoglobin, the oxygen acts as a Lewis base, furnishing a lone pair of electrons for a coordinate covalent bond.

97. (B) Formal charge (FC) = # of valence electrons − (# of nonbonding electrons + ½ # of shared electrons). For the hydrogen, $1 − (0 + 1) = 0$. There is no need to calculate the other atoms, since answer B is the only one that contains a zero in the first position.

98. (D) The left oxygen has two bonds and two lone pairs in a tetrahedral orientation. That is sp^3 hybridization. The right oxygen has a double bond, which is sp^2 hybridization.

99. (A) When a hydrogen ion is lost, the nitrite ion can be written in two possible resonance forms with the nitrogen to oxygen bonds switched. Therefore, in the resonance hybrid, the bonds are equal in length, with a bond order of 1.5.

100. (C) Both carbon atoms have three groups attached. (Remember that a double bond counts as a single group.) The geometry with three groups around a central atom is trigonal planar.

101. (B) The formula of the phosphate ion is PO_4^{3-}. There are four groups surrounding the phosphorus atom and no lone pairs. Therefore, the molecular geometry is tetrahedral.

102. (D) This is an example of an acid-base reaction. A hydrogen ion transfers from the HCl to the ammonia, forming ammonium cations and chloride anions, which form an ionic compound. Most ionic compounds are solids at room temperature.

103. (A) Both ethyl alcohol and carbon monoxide are molecular compounds that do not ionize in water. They are not electrolytes. HBr will ionize, producing hydrogen ions and bromide ions, and is an electrolyte.

104. (D) Free radicals contain an unpaired electron. Any species with an odd number of electrons is a free radical. Otherwise, it would be necessary to draw Lewis structures or MO diagrams. In drawing the Lewis structures for answers A, B, and C, all the electrons are paired. However, in NO_2, there are an odd number of valence electrons and an unpaired electron, which is normally shown on the nitrogen atom.

105. (C) sp^3 hybridization is tetrahedral (109.5°), sp is linear (180°), and sp^2 is trigonal planar (120°).

106. (C) The chlorate ion would have the chlorine atom bonded to three oxygen atoms, and it would also have a lone pair of electrons, making its molecular geometry trigonal pyramidal.

107. (B) Silicon tetrafluoride would have the same geometry as carbon tetrachloride (tetrahedral). It would be a covalent compound (nonmetal + nonmetal). There are no common reactions that would cause rapid oxidation of SiF_4, resulting in an explosion. Because it is tetrahedral, it would be nonpolar.

108. (A) In nitric acid, the hydrogen is bonded to one of the oxygen atoms. In order for oxygen to have a complete octet, there must be a double bond (4 electrons) and two single bonds (2 × 2) for a total of 8 electrons.

109. (D) Neither calcium ions nor ethane have nonbonding electron pairs, eliminating answers A and C. Answer B is false, since the presence of the electron pairs would reduce any positive charge. A Lewis base is defined as an electron pair donor, and that is what ligands do, making answer D correct.

110. (B) Because of its size (it is the smallest of the halogens), it is difficult to fit four oxygen atoms around it, and it cannot form a +7 oxidation state. Again, because of its size, it is impossible for it to exceed an octet of electrons.

111. (A) Hydrogen bonding in water is a relatively strong intermolecular force. It is stronger than the hydrogen bonding that occurs in ammonia, since oxygen is more electronegative than nitrogen. Water's hydrogen bonding is stronger than the dipole-dipole forces in CH_3F and much stronger than the London dispersion forces in CH_3CH_3. It is, however, weaker than the ionic bonding that is in sodium chloride.

112. (D) The $H-N-H$ bond angle is less than predicted because the lone pair of electrons on the nitrogen spread out slightly, forcing the bond angle to be a little smaller.

113. (A) In the Lewis structure of the cyanate ion, there are two double bonds, one to the oxygen and one to the nitrogen. There are no lone pairs in the carbon, so there are two groups around the carbon, giving it a linear geometry.

114. (B) The level is always read at the bottom of the meniscus. Many times, it is difficult to tell exactly where the top of the meniscus is located.

115. (C) The nitrite ion, NO_2^-, has a single bond to one oxygen atom, a double bond to the other oxygen atom, and a lone pair of electrons on the nitrogen atom. This results in a bent molecular geometry. Resonance does not alter this.

116. (D) The electromagnetic spectrum in order of decreasing energy is: gamma ray → x-ray → ultraviolet → visible → infrared → microwave → radio waves. Since the vibrational transitions require less energy, they will be to the right of the ultraviolet region. The only choice to the right is infrared.

117. (A) Xe is monatomic with no vibrational spectrum, since there are no bonds. All the others are molecules with vibrational spectra due to their bonds.

118. (D) Answer A is wrong, since carbon dioxide has double bonds. Answers B and C are basically the same, and therefore, neither can be correct. The correct answer is D. The presence of polar covalent bonds allows energy to be absorbed. Remember that a molecule can have polar bonds but not have a permanent dipole.

119. (B) Since there are no hydrogen atoms bonded to the fluorine, there can be no hydrogen bonding. London dispersion forces are very weak. Covalent bonding is an intramolecular force, not an intermolecular force. CHF_2Cl would have a dipole, and so there could be dipole-dipole interaction.

120. (A) The geometry of PF_3 should be similar to that of NH_3, which is trigonal pyramidal. This eliminates answers C and D. Formal charge (FC) = # of valence electrons − (# of nonbonding electrons + ½ # of shared electrons). Phosphorus has 5 valence electrons, so it would have the following formal charge: FC = 5 − [2 + ½(6)] = 0.

121. (A) All three of these elements are adjacent in the same period. Ionization energies increase with decreasing atomic radii, which occurs in a period as the effective nuclear charge increases. Therefore, cadmium would be predicted to have the highest ionization energy, followed by indium, with tin having the smallest.

122. (D) Metals tend to be malleable, ductile, and good conductors; they are not brittle. Nonmetals tend to be brittle.

123. (A) The boron halide has a vacant orbital that can accept an electron pair. That makes it a Lewis acid. BX_3 would have a trigonal planar geometry (only three electron pairs around the boron), but it changes to tetrahedral when the fourth electron pair is accepted.

124. (A) Within a period, electronegativities increase as you move from left to right on the periodic table. In the F—F bond, there would be a zero polarity difference, since the electronegativities are the same. This eliminates answers C and D. The greatest will be between C and F, since the difference in electronegativities is the greatest. Therefore, answer A is correct.

125. (C) Because of the properties of diamond (hardness, etc.), the bonds between the carbons must be strong. This eliminates answers A and D, because these are weak forces. The bond between Zn and S is ionic (metal + nonmetal), but the bond between carbon atoms must be covalent (nonmetal + nonmetal).

126. (B) The stronger the intermolecular forces, the higher their heat of vaporization. Water's intermolecular force is hydrogen bonding; HCl's would be dipole-dipole attraction, while chlorine's would be London forces. All of these are weak relative to the ionic bonding present in NaCl.

127. (A) The molar mass of HCl is 36.5 g/mol. Therefore, 73 g would be 2.0 moles, but since it is a strong acid, you would get 4.0 moles of particles. Since freezing point depression is a colligative property, it just depends on the number of particles present. If 2 moles of particles lowered the freezing point 3.8°C, then 4 moles should lower it 7.6°C from pure water's freezing point of 0°C. Therefore, the freezing point of the solution should be −7.6°C.

128. (C) Ionization energies, the energy needed to remove a valence electron from an atom, increase from left to right in a period and decrease from top to bottom in a group. The element with the lowest ionization energy should then be in the leftmost group in the period and the lowest on the periodic table. That would be potassium in this case.

129. (C) Both electronegativity and electron affinity are related to the increasing effective nuclear charge, which occurs with increasing atomic number within a period. In general, as the atomic number increases, the atoms become smaller and the attractive force for the valence electrons increases (increasing electronegativity). This eliminates answers A and B. The electron affinity generally increases with increasing atomic number within a period, since the increasing effective nuclear charge should attract an electron being added to the valence shell. This eliminates answer D.

130. (A) Linear, tetrahedral, and octahedral are all symmetrical geometries and thus can be nonpolar. Bent molecules are not symmetrical and cannot be nonpolar.

131. (C) Sodium, because it is the leftmost member of its period, has a low ionization potential. It is relatively easy for it to lose an electron, forming a cation and reaching noble gas configuration.

132. (B) Argon is a noble gas and therefore is very difficult to react. All of the other choices will react with potassium.

133. (B) Answer A is false because chloride ions are much larger than sodium ions. Answer C is false since calcium ions are smaller than potassium ions. Answer D is false since potassium ions would be larger than sodium ions. Answer B is correct since the sodium ion would be smaller than the potassium ion, given that potassium has added a major energy level.

134. (C) Neon is a noble gas with a full octet. It would require a tremendous amount of energy to add an additional electron. All the rest are nonmetals that readily form anions.

135. (C) The sulfide ion is S^{2-}. It has gained two electrons and now has the same number of electrons as argon, Ar.

136. (D) Trigonal bipryamidal has five groups around the central atom; square planar has four and two lone pairs; tetrahedral has four groups; and octahedral has six groups and no lone pairs.

137. (A) Residues refer to the group (side chain) attached to the amino acid backbone—amine-carbon-carboxylic acid. In answers B, C, and D, the residues are either carbon or hydrogen atoms that have no nonbonding electron pairs. However, aspartic has a residue that contains two oxygen atoms that have nonbonding electron pairs that could form coordinate covalent bonds with an iron.

138. (A) Formal charge (FC) = # of valence electrons − (# of nonbonding electrons + ½ # of shared electrons). Nitrogen has 5 valence electrons, so it would have the following formal charges in the answers—answer A: FC = 5 − (2 + 3) = 0; answer B: FC = 5 − (6 + 1) = −2; answer C: FC = 5 − (4 + 2) = −1; answer D: FC = 5 − (6 + 1) = −2.

139. (A) Lowering the pH will add a hydrogen ion to the pyruvate ion, making it neutral. This will decrease its solubility in water. This eliminates answer B. In the ion shown, the carbon-oxygen bonds on the left of the diagram participate in resonance, which makes the bonds of equal length and longer than the bond to the upper right oxygen atom. This eliminates answers C and D.

140. (D) A trigonal pyramidal geometry has three atoms, and one lone pair surround a central atom. Sulfur trioxide is trigonal planar, water is bent, and methane is tetrahedral. Ammonia has three hydrogen atoms and a lone pair of electrons around the nitrogen; it is trigonal pyramidal.

141. (A) The reaction is $CO_2(g) + H_2O \rightarrow H_2CO_3$. The Lewis structure of carbon dioxide has two double bonds to the oxygen atoms, making it linear. This eliminates answers B and D. In carbonic acid, the hydrogen atoms are attached to oxygen atoms, making two of the oxygen atoms single bonded to the carbon and one doubled bonded. That gives three groups around the carbon and a trigonal planar geometry.

142. (A) Cell membranes are hydrophobic, allowing the passage of nonpolar materials more easily than polar or ionic ones. Since dimethyl mercury is nonpolar, it can pass through the cell membranes more easily than the cationic mercury.

143. (C) Melting points can be correlated with molar mass and intermolecular forces. The higher the molar mass, the higher the melting point. All molecules except answer C can participate in hydrogen bonding. The lower intermolecular attraction in answer C leads to a lower melting point. Answer C also has the lowest molar mass.

Chapter 4: Phases

144. (B) When the vapor pressure of a liquid equals or exceeds the external pressure, it boils.

145. (B) Denver is at a much higher altitude than Houston is and has a lower atmospheric pressure. It takes less energy to heat the water to a point where its vapor pressure exceeds atmospheric pressure and it begins to boil.

146. (A) The heat of vaporization is related to the strength of the intermolecular forces in the liquid. The stronger the intermolecular forces, the higher the heat of vaporization. Water exhibits hydrogen bonding, a strong intermolecular force, between molecules. It takes a lot of energy to break these strong hydrogen bonds between the molecules, so the heat of vaporization is high.

147. (C) Sublimation is a process that starts with a solid, not a gas. Evaporation and fusion are processes that start with liquids, not gas. Deposition starts with a gas, and if the temperature decreases at constant pressure, the gas could go from a gas to a solid (deposition).

148. (D) Normal boiling point refers to the temperature at which a liquid will boil at 1 atm pressure. If the pressure is increased, the liquid will boil at a temperature greater than its normal boiling point.

149. (A) Normal boiling point refers to the temperature at which a liquid will boil at 1 atm pressure. Answer D, STP, assumes a temperature of 0°C. Answer B is wrong because changing the pressure affects the boiling point, and answer C has nothing to do with boiling point.

150. (C) Graphite is softer than diamond because there are relatively weak London dispersion forces between the sheets of graphite, while in a diamond, there are strong covalent bonds throughout the structure. The stronger the bonds holding the entire structure together, the harder the material.

151. (A) The heat of vaporization is related to the strength of the intermolecular forces. Hydrogen sulfide and formaldehyde have dipole-dipole attraction, propane would have London dispersion forces, and ethanol would have hydrogen bonding. Hydrogen bonding is the strongest of these; therefore, ethanol would have the highest heat of vaporization.

152. (A) Answer B is incorrect because the pressure is due to a variety of gases, not just ozone. Answer C is incorrect because the boiling point of water varies with atmospheric pressure. Answer D is incorrect because the energy of collisions is not affected by pressure, just temperature. Answer A is correct because at pressures lower than 1 atm (760 torr), water will boil at a lower temperature. (It takes less energy for the water molecules to exceed atmospheric pressure.)

153. (D) In order for a liquid to freeze, energy must be lost to the surroundings. In deposition, going from a gas to a solid, energy must be lost to the surroundings. The same is true for condensation, going from a gas to a liquid. However, in sublimation, in order for a solid to go into the gaseous phase, energy must be absorbed from the surroundings.

154. (C) Heat capacity has units of J/K, while the specific heat has units of J/g·K. Multiplying the specific heat by the mass (g) of the substance generates the heat capacity.

155. (C) Answers A, B, and D represent phase transitions that would be represented as horizontal portions on the graph. The specific heats are necessary to calculate the heat needed to go from one temperature to another.

156. (D) Fusion plus vaporization represents the process of going from a solid to a gas; this is sublimation.

157. (A) The melting point occurs at T_2, and boiling occurs at T_3. These temperatures must be known in order to calculate the temperature change from the liquid at the melting point to the liquid at the boiling point.

158. (B) 36 g of ice is 2.0 moles of ice. To completely melt all the ice, it would take (6.0 kJ/mol × 2 mol) 12.0 kJ. Only 6.0 kJ is being added. Therefore, not all of the ice would melt, and the temperature would remain constant at 0°C.

159. (C) The critical point is the point at which the gas and liquid phases are indistinguishable from each other. The critical temperature would be T_2, and the critical pressure would be P_2.

160. (A) The heats of fusion of ionic compounds are normally much higher than those of covalent substances, even polar ones. The heat of fusion of aluminum oxide will be higher than that of NaCl because of the increased attraction of the 3+ cations and 2− anions versus the 1+ to 1− attraction in sodium chloride. These facts eliminate answers B, C, and D. (The heat of fusion of a polar covalent substance [H_2O] should be greater than that of a nonpolar one [CCl_4].)

161. (C) NaCl is a nonvolatile solute, so it would not appear in the gas phase. Water does not decompose upon boiling, so only water vapor is in the gas phase.

162. (C) The water is initially at T_1, and the temperature remains constant. The water vapor is compressed, which causes the pressure to increase along the vertical dotted line. It goes from a gas to a solid and then to a liquid.

163. (A) The heat of sublimation is the heat necessary to both melt and vaporize a substance. With both graphite and diamond, extremely strong covalent bonds must be broken, requiring a large amount of energy. Answers B, C, and D are incorrect because they do not mention covalent bonds.

164. (C) This is a Hess law problem. Looking at the desired reaction, you can see that you need 2 N_2 on the left, so you can multiply the third equation by 2 along with its ΔH (giving −700 kJ). You need 2 N_2O_5 on the right, so reverse equation two and multiply by 2 (giving +160 kJ). Finally, since you will need 2 H_2O on the left to cancel the 2 H_2O that are now on the right, reverse equation one (giving + 570 kJ). Summing up the three ΔH values, you get 30 kJ.

165. (B) Specific heat has the units of J/g K. In this case, there was a 60°C change in temperature, which is also a 60 K change in temperature (no need to convert from °C to K). Therefore, the specific heat would be 240 J / (40 g × 60 K) = 0.10 J/g·K.

166. (A) The molar mass of ethanol is 46 g/mol, so 0.46 g of ethanol is 0.01 moles. The heat absorbed by the calorimeter would be 1.2 kJ/°C × 10°C = 12 kJ, and it would be negative, since the reaction was exothermic. The heat of reaction would be −12 kJ / 0.01 mol = −1.2 × 10^3 kJ.

167. (B) The heat change would be the specific heat × mass of solution × change in temperature = 4.2 J/g·K × 158 g × 3.0 K = 1.99 × 10^3 J, or about 2 kJ. This enthalpy change was for 0.10 moles of sodium nitrate (8.6 g / 86 g/mol), so the enthalpy change would then be 2 kJ / 0.10 mol = 20 kJ/mol, and the value would be positive, since the reaction was endothermic.

168. (A) 260 g of KI is about 2 moles. The total energy change will be KI*(s)* → KI*(l)* + heating KI*(l)* from −50°C to −35°C + KI*(l)* → KI*(g)* + heating −35°C to 0°C: [2 mol × 2.9 kJ/mol] + [2 mol × 44 J/mol·K × 15 K] + [2 mol × 44 kJ/mol] + [2 mol × 25 J/mol·K × 35 K] = 97 kJ. Appropriate approximations should be applied to avoid arithmetic.

169. (C) Remember in thermochemistry problems, it is products minus reactants. The heats of formation of the elements, potassium and hydrogen, are zero. Therefore, $2(\Delta H_f^\circ \text{ of KOH}) - 2(\Delta H_f^\circ \text{ of H}_2\text{O}) = -130$ kJ. Then substitute the standard heat of formation of water and solve for the standard heat of formation of KOH, which equals −350 kJ/mol.

170. (D) At the top of the mountain, the atmospheric pressure is lower, and less heat is necessary to heat the water enough so that the water's vapor pressure is equal to the atmospheric pressure. Therefore, the water boils at a lower temperature.

171. (A) Water exhibits hydrogen bonding between the molecules. When water begins to freeze, a crystal structure is formed using those hydrogen bonds. In this crystal structure, there are large empty spaces that make the density of ice less than the density of water.

172. (C) The heat of vaporization is related to the strength of the intermolecular forces in the liquid. The stronger the intermolecular forces, the higher the heat of vaporization. Water exhibits hydrogen bonding between molecules, while methyl chloride has dipole-dipole attraction as its predominant intermolecular force. Hydrogen bonding is a stronger force than dipole-dipole attraction, so water has a higher heat of vaporization.

173. (B) At the boiling point, the energy needed to vaporize the liquid is the heat of vaporization times the mass of the substance. Therefore, 0.500 kJ/g × 100.0 g = 50.0 kJ.

174. (B) The only time you can add energy to a sample and have the temperature remain constant is at a phase transition. The problem states that the ethanol is at a temperature below its boiling point. The only other phase transition would be melting, so the sample is at its melting point.

175. (B) Since energy must be absorbed to boil a liquid, the enthalpy change is positive. This eliminates answers C and D. This energy is used to break the forces between the particles, the intermolecular (not intramolecular) forces.

176. (A) To lower the freezing point of the water and prevent the sidewalk from cracking, a substance is added to make a solution of a lower freezing point. This freezing point depression is a colligative property, so it really depends on the number of solute particles present. Answers B, C, and D involve salts, so for every mole of solute added, you get 2 to 4 moles of particle. Answer A, however, is a molecular substance, so for every mole added, you get only 1 mole of particles; it will have the least effect.

177. (B) At the pressure given and $-10°C$, the substance would be in the solid phase. Increasing the temperature at that pressure would cause the substance to move from the solid to the gaseous phase. This process is sublimation.

178. (B) Deposition is going from the gaseous phase to the solid phase: $SO_3(g) \rightarrow SO_3(s)$. It would be the difference in the enthalpies of formation of products minus reactants, or -454 kJ/mol $- (-396$ kJ/mol$) = -58$ kJ/mol.

179. (A) The first horizontal region represents melting. The second horizontal region represents vaporization: going from a liquid to a gas. This eliminates answers D and C. Normally, it takes more energy to convert a substance into a gas than to melt a solid, so that the heat of vaporization is greater than the heat of fusion.

180. (B) The heat capacity is the specific heat of the substance times its mass. It has units of J/°C. The units eliminate answers A and C. 4.2 J/g°C × 20 = heat capacity = 84 J/°C.

181. (A) The enthalpy change would be the specific heat of solution times the mass (solute + water) of the solution times the change in temperature. It would be a negative value, since the process was exothermic (that eliminates answers B and D). Enthalpy change = 4.2 J/g°C × 200 g × 10.0°C = -8.4 kJ.

182. (D) The critical point is the point at which the gas and liquid phases are indistinguishable from each other.

Chapter 5: Gases

183. (D) Dalton's law states that the sum of all partial pressures of gases in a mixture equals the total pressure. This means that the partial pressure of a gas is the percentage (or mole fraction) of the gas times the total pressure. In this case, (35%/100%) × 1,000 torr = 350 torr.

184. (B) Pressure is related to the collisions of the gas particles with the inside walls of the container. Increasing the number of moles of gas, at constant temperature, would cause the pressure to increase.

185. (C) Gases that are at the same temperature have the same average kinetic energies.

186. (D) In order for a shock absorber to work, the material must be able to compress to absorb the shocks. Gases are compressible, so shock absorbers are normally filled with air or nitrogen gas. Water would not work because it is not compressible. Answer A is incorrect since you would not be exerting enough pressure in a shock absorber; Answer B would be correct, but shock absorbers must work at lower temperatures; therefore, answer D is a better reason. Answer C has nothing to do with shock absorbing ability.

187. (C) Consider the ideal gas equation, $PV = nRT$. Solving for P: $P = nRT/V$. If the pressure doubles and the volume stays constant, the temperature must double (R is a constant, and the number of moles remains constant).

188. (D) The diffusion of gases (Graham's law) relates to the velocity (v, not V) of the gas particles. Answer A is the modification of the ideal gas law for real gases, answer B gives the pressure/volume relationship of gases, and answer C is the ideal gas equation.

189. (C) Since the cloud formed near the center, the molar mass of the unknown is close to that of dimethylamine, 45 g/mol. HI = 128 g/mol, HF = 20 g/mol, HCOOH = 46 g/mol, and HSCN = 59 g/mol. HCOOH is the closest.

190. (B) The most ideal will be the one with the weakest intermolecular forces. HF is highly polar, so it is eliminated. The remaining three have London dispersion forces. Helium is the smallest (fewest electrons); therefore, it has the weakest London dispersion forces, which leads to the minimum deviation.

191. (C) Since the pressure was constant, the only effect will be the change in temperature from 273 K to 546 K ($\times$ 2). The initial volume was 11.2 L, and the volume should increase with an increase in temperature. The new volume will be 11.2 L $\times$ 2 = 22.4 L.

192. (C) The piston was locked into place, so the initial and final volumes are the same, eliminating answers A and D. The temperature was doubled (273 K to 546 K), so at constant volume the pressure should double, eliminating answer B. Answer C is correct with $P_f = 2 P_i$ and the two volumes equal.

193. (A) The gas is heated to 546 K, and then the piston is unlocked. The gas will expand, $V_f > V_i$, and as it expands, the pressure will decrease, since the volume is greater, $P_f < P_i$.

194. (A) 76 torr = 0.1 atm, and (10,000 Pa) / (1 × 10⁵ Pa/atm) = 0.1 atm, so the total pressure is (0.50 + 0.10 = 0.10) = 0.70 atm.

195. (B) Converting the masses to moles gives CH_4 = 1.0 mole, C_2H_6 = 1.0 mole, and C_4H_{10} = 1.0 mole. The mole fraction of each, times the total pressure, will give the partial pressure of each gas. (1 mol/3 mol) × 1.20 atm = 0.40 atm for each gas.

196. (A) 1.0 mole of any gas at 1 atm and 273 K (0°C) = 22.4 L. Therefore, there would be 22.4 L × 2.0 g/L = 44.8 g in 1.0 mol. The molar masses are CO_2 = 44 g/mol, CH_4 = 16 g/mol, H_2 = 2.0 g/mol, and Xe = 131 g/mol. The gas must be carbon dioxide.

197. (B) The volume and Kelvin temperature of a gas are directly related at constant pressure and amount. The gas began at a temperature of 298 K. The volume was reduced by one-half; therefore, the temperature would decrease by one-half. The new temperature would be 298 K/2 = 149 K = −124°C. Answer B is the closest.

198. (C) Samples of gases that are at the same volume, temperature, and pressure have the same number of gas particles.

199. (A) If you have two samples of gas at the same temperature, they have the same average kinetic energies. However, the hydrogen gas, because it has a smaller molar mass, would have a higher average velocity.

200. (D) Graham's law states that the rate of effusion of a gas is inversely proportional to the square root of its molar mass. In comparing two gases, it would be r_B/r_A = $(M_A/M_B)^{1/2}$. Assume that gas A is methane (molar mass of 16 g/mol) and has a relative rate of effusion of 1. The unknown would have an effusion rate of 0.5. Substituting those values into the equation gives 0.5/1 = (16 g/mol/M_B)$^{1/2}$. Solving for M_B gives 64 g/mol. Molar mass of oxygen is 32 g/mol, helium is 4 g/mol, carbon dioxide is 44 g/mol, and sulfur dioxide is 64 g/mol. The best choice is sulfur dioxide.

201. (C) The gas closest to ideal is oxygen because its van der Waals constants are the smallest.

202. (B) At high pressure, there are many gas molecules colliding with the inside walls of the container. This means there are a large number of molecules, which makes the volume of the molecules themselves no longer negligible.

203. (A) Answers C and D are valid at low temperatures. Answer B, concerning the volume of the gas particles being negligible, is related to the pressure. Answer A, the interaction between particles, becomes more significant when the kinetic energies of the particles are lower (at lower temperatures).

204. (C) Reducing the pressure by one-half would cause the volume to double. Doubling the absolute temperature would also cause the temperature to double, making the two volumes the same. (Answers B and D are the same answer; therefore, neither can be correct.)

205. (A) The ideal gas equation assumes ideal behavior on the part of the gas. With water vapor, there is a strong interaction between the polar gas molecules, hydrogen bonding, which causes the deviation in pressure from that predicted by the ideal gas equation.

206. (C) 1 mole of any ideal gas at STP occupies 22.4 L, so 2 moles would occupy 44.8 L at STP.

207. (C) Dalton's law says that the sum of the partial pressures of gases in a gas mixture equals the total pressure. 35 torr + 15 torr + P_{Ar} = 100 torr. The partial pressure of Ar is 50 torr. The percentage of Ar would be (50 torr/100 torr) $\times$ 100% = 50%.

208. (D) Graham's law states that the rate of effusion of a gas is inversely proportional to the square root of its molar mass. Rounding is very important. In comparing two gases, it would be $r_B/r_A = (M_A/M_B)^{1/2}$. For answer A, Kr and He, it would be $[(80 \text{ g/mol}) / (4.0 \text{ g/mol})]^{1/2} > 2$ (incorrect); for answer B, O_2 and N_2, it would be $[(32 \text{ g/mol}) / (28 \text{ g/mol})]^{1/2} \neq 1$ (incorrect); for answer C, H_2 and Xe, it would be $[(2 \text{ g/mol}) / (130 \text{ g/mol})]^{1/2} < 2$ (incorrect); for answer D, it would be $[(80 \text{ g/mol}) / (20 \text{ g/mol})]^{1/2} = 2.0$ (correct).

209. (A) The latex balloon contains pores that allow gases to effuse through the latex. Since the concentration (and pressure) of gases inside the balloon is greater than outside the balloon, effusion will take place from inside to outside. Graham's law states that the lower the molar mass, the faster the rate of effusion, so the helium will effuse out faster than the heavier gases that comprise the air. After some helium has effused out, the balloon will not rise as high as before.

210. (A) Halving the diameter of the piston will cause the volume to decrease to one-half its former volume; doubling the pressure will cause the piston to move outward, doubling its initial volume. The result of the two processes is no net change in volume.

211. (A) This is a Graham's law problem. The lower the molecular mass (weight), the farther it will travel. Since HX1 traveled the least distance, it has the highest molecular weight.

212. (A) The system initially is at STP, so the 0.5 moles of gas would occupy 11.2 L. (1 mole of any gas at STP occupies 22.4 L.) Now you can convert the volume to cubic meters: 11.2 L $\times$ (1,000 mL/1 L) $\times$ (1 cm^3/1 mL) $\times$ (1 m)^3/(100 cm)^3 = 1.1 $\times$ 10^{-2} m^3.

213. (D) Kinetic molecular theory assumes that the intermolecular forces between gas particles are negligible (eliminating answer A), that the average kinetic energies are all the same (eliminating answer B), and that the volumes of the gas particles are negligible (eliminating C). It does not assume that the average velocities are the same.

214. (C) The sum of the mole fractions must equal 1, so this eliminates answers A and B. The mole fraction is independent of the volume and temperature, so this eliminates answer D. The mole fractions would be the individual gas pressure divided by the total pressure, which is 1.0 atm. So for He, it would be 0.50 atm/1.0 atm = 0.5; Ne: 0.20 atm/1.0 atm = 0.20; and Ar: 0.30 atm/1.0 atm = 0.30.

215. (C) You can use the ideal gas equation, $PV = nRT$, and solve for V (with rounding): $V = nRT/P = [(1.0 \text{ mol})(0.08206 \text{ L·atm/K·mol})(350 \text{ K})]/1.0 \text{ atm} = [(1.0 \text{ mol}) (0.1 \text{ L·atm/K·mol})(350 \text{ K})] / 1.0 \text{ atm} = 28.7 \text{ L}$. Using 0.1 instead of 0.08206 for the ideal gas constant gives a rounded value of 35 L; 28.7 L is closest.

216. (C) Use the ideal gas equation and solve for n (with rounding): $n = PV/RT = (3.0 \text{ atm})(0.080 \text{ L}) / (0.08 \text{ L·atm/K·mol})(300 \text{ K}) = 10^{-2}$ moles. Therefore, $0.6 \text{ g}/10^{-2} \text{ mol} = 60 \text{ g/mol}$. The molar masses are $CCl_4 = 154$ g/mol, $C_8H_{18} = 114$ g/mol, $C_4H_{10} = 58$ g/mol, and $CH_3OH = 32$ g/mol. C_4H_{10} will give a molar mass closest to the calculated value.

217. (A) Whenever a gas is collected over water, there will be water vapor mixed in with the gas. This increases the total pressure of the gas collected.

Chapter 6: Solutions

218. (A) When a gas is collected over water, there will be some water vapor mixed in the gas. This comes from evaporation of the water and establishment of the water's vapor pressure.

219. (A) HNO_2 is a weak acid, while $NaNO_2$ is a strong electrolyte and will completely dissociate. If you add an equal number of moles of each to water, there will be more ions in the sodium nitrite solution than in the nitrous acid solution. Since the amount of freezing point depression depends on the number of solute particles, you will have a greater depression of the freezing point with the sodium nitrite solution. Answers C and D involve factors that have nothing to do with the freezing point depression.

220. (A) A solution containing a nonvolatile solute always has a vapor pressure lower than that of the solvent because some of the solute particles will be at the surface, reducing the number of particles that can escape as vapor.

221. (A) Sulfur dioxide is very soluble in a solution of NaOH because it reacts with the NaOH as shown by the following reaction: $SO_2 + 2\,NaOH \rightarrow Na_2SO_3 + H_2O$.

222. (D) By the solubility rules, answers A, B, and C would be very sparingly soluble, while calcium acetate should be very soluble.

223. (A) The vapor pressure of water is affected by the temperature (the higher the temperature, the higher the vapor pressure), by hydrogen bonding (an important factor in determining vapor pressure for water), and by the presence of solutes (solutions of nonvolatile solutes have a lower vapor pressure than the pure solvent). The external pressure does not affect the vapor pressure.

224. (B) When the HCl dissolves in the water, it ionizes completely (eliminating answer A), releasing hydrogen ions, which lowers the pH (eliminating answer C). Since a solution has been formed, the boiling point rises. (The boiling point of a solution is always higher than that of the pure solvent unless there is a very volatile solute present.)

225. (A) Decanoic acid is a carboxylic acid that will react with a base. The anion formed will have a hydrophilic end, the carboxylate end.

226. (D) The solution that boils at the highest temperature will be the one with the greatest number of solute particles. Nitric acid is a strong acid, so a 1.0 M nitric acid would contain 2.0 M in ions. The same is true of the 1.0 M NaCl. Ethyl alcohol is a nonelectrolyte, so it would form a solution 2.0 M in particles. Calcium chloride is a strong electrolyte and would yield a solution 3.0 M in ions.

227. (B) A solution is a homogeneous mixture and will have the same concentration throughout. The solution will have the same freezing point whether the whole solution is tested or only a portion of the solution.

228. (D) The freezing point depression is a result of the lowering of the vapor pressure of the solution.

229. (C) Osmosis and osmotic pressure are colligative properties and are dependent on the number of solute particles present. In answer A, the NaCl is a strong electrolyte, but the acetic acid is a weak electrolyte. Therefore, the solutions will have different numbers of particles and different osmotic pressures. If a semipermeable membrane separates two solutions having the same osmotic pressures, the solvent can pass through in both directions (eliminating answer B). Osmotic pressure is the pressure needed to just stop osmosis (eliminating answer D). In osmosis, the solvent always passes from the solution of lower concentration to one of greater concentration.

230. (A) Deviation from the predicted van't Hoff factor occurs when solute particles are close together and they can interact, forming ion pairs. The least variance can be expected when the solution is very dilute.

231. (D) In a 0.1 m solution of sodium nitrate, there will be some ion pairing, making the actual van't Hoff factor less than the ideal one. This means that the amount of freezing point depression would be less than expected and the actual freezing point should be higher than the theoretical freezing point that was expected.

232. (A) The use of colligative property measurements (FP depression, BP elevation, and osmotic pressure) is an effective way of measuring the van't Hoff factor since it appears in their equations. However, density measurements do not involve the van't Hoff factor.

233. (B) Magnesium and calcium ions, because of their 2+ charge, would have a greater affinity for the resin and would exchange well. Li^+ would be attracted well to the resin because its small size gives it a higher charge density than sodium ions. Potassium ions are larger than sodium ions with the same 1+ charge, so their charge density is less, and they therefore have less affinity for the resin than sodium.

234. (A) The sodium, calcium, and aluminum ions are all very stable and difficult to reduce, but Hg^{2+} can be reduced to Hg_2^{2+} or Hg.

235. (D) Compounds A, B, and C are all covalent molecules and should be somewhat volatile. However, $HgCl_2$ is ionic and will not have an appreciable vapor pressure in solution.

236. (B) Adding a low-volatility solute will decrease the vapor pressure (vapor pressure lowering is a colligative property). This eliminates answer A. The amount of lowering will be relatively small, not a large amount as in answers C and D.

237. (B) This is a freezing point depression problem. The solution with the greatest number of particles will have the lowest freezing point. Sodium sulfide (answer D) yields 3 particles, sodium phosphate (answer C) yields 4 particles, and sodium nitrate (answer B) and sodium chloride (answer A) both yield 2 particles. The sodium phosphate solution will freeze last (the most particles).

238. (A) The precipitate Ag_3PO_4 will have the strongest ionic bond because of the 3− charge on the phosphate ion giving a greater attractive force to the silver cations. Therefore, the Na_3PO_4 solution (answer A) is correct.

239. (A) The reaction would be NaOH*(aq)* + HNO$_3$*(aq)* → NaNO$_3$*(aq)* + H$_2$O*(l)*. There are equal numbers of moles of OH$^-$ and H$^+$, so the only thing left in the solution will be water and a NaNO$_3$ solution. Since the concentrations were both 4.00 *M* and equal volumes were mixed (doubled), the concentration of the NaNO$_3$ solution would be 2.00 *M*. Since the solution is not as concentrated as both original solutions, it would not freeze at as low a temperature as the original solutions (colligative properties).

240. (D) A more useful representation of the original formula would be [Pt(NH$_3$)$_4$]Cl$_2$, which dissociates upon dissolving to [Pt(NH$_3$)$_4$]$^{2+}$ plus 2 Cl$^-$, for a total of three ions. The [Pt(NH$_3$)$_4$]$^{2+}$ ion, like most four-coordinate platinum(II) complexes, is square planar. Answer A cannot be correct, because there are too many ligands present. Answer B, while also four-coordinate, does not normally occur for platinum. If the compound were octahedral, there would be no ions released upon dissolution, which eliminates answer C.

241. (A) Calcium carbonate will react with the acid to form carbon dioxide and water. This will make the solubility of the calcium carbonate in the 1.0 *M* HCl greater than that of the other salts, which do not react.

242. (A) The solubility in water is related to having a polar end (all do), but the one with the smallest nonpolar end will be the most soluble (answer A).

243. (B) Vapor pressure, freezing point, and boiling point are all colligative properties; that is, they depend on the number of particles present. Density does not depend on the amount present: a big piece of gold has the same density as a small piece of gold.

244. (D) The formula of magnesium chloride is MgCl$_2$. It is a strong electrolyte, so that 3 moles of ions are formed from 1 mole of the salt. The solution is 2 molal, or 2 moles of salt per kg solvent; 2 moles of salt will yield 6 moles of ions, so the solution is 6 *m* in particles. The boiling point elevation can be calculated by (0.5°C/*m*) × 6 *m* = 3°C. That is the amount of elevation; the actual boiling point would be 100°C + 3°C = 103°C.

245. (B) The freezing point is affected by the number of solute particles present. 1 mole of sodium nitrate yields 2 moles of ions, while 1 mole of calcium nitrate yields 3 moles of ions. Therefore, the calcium nitrate solution will exhibit more freezing point depression and will freeze at a lower temperature.

246. (C) Solubility is affected by the natures of the solute and solvent: "Like dissolves like." This means that water, a polar solvent, will tend to dissolve polar solvents and not nonpolar ones. Oxygen is nonpolar, and the others are polar.

247. (B) An ideal solution is one that obeys Raoult's law. The more dilute the solution is in terms of particles, the more ideal. Both answers A and C would yield 0.03 M in particles, because you have 3 moles of ions per 1 mole of salt. Answer D would yield a solution that is 0.05 M in ions, but the NaCl solution would yield a solution only 0.02 M in ions.

248. (A) Notice that answers A and B conflict; therefore, one of the two must be the correct answer. There is no need to waste time with the other two answers. Since the vapor pressure of liquid A is higher than that of liquid B, it will take more energy to raise liquid A to the point where its vapor pressure exceeds atmospheric pressure, and thus, its boiling point is higher.

249. (A) The pH will not have a significant effect on the gas exchange, and neither will volume changes in the lungs. The temperature will affect both equally, so it will not have an effect. The concentration gradient will have an effect, as the passage of the gases will be from an area of higher concentration to one of lower.

250. (C) The boiling point is a colligative property. The greater the concentration of solute particles, the higher the boiling point. Answers A, B, and C are all soluble salts, while D is a nonelectrolyte. A 2.0 m solution of ethyl alcohol (answer D) will be 2.0 m in particles; 1.0 m sodium nitrate will be 2.0 m in ions; 1.0 m calcium chloride will be 3.0 m in ions; and a 1.0 m solution of aluminum nitrate will be 4.0 m in ions. Since the aluminum nitrate solution has the highest concentration of ions, it will have the highest boiling point.

251. (C) In the KCl solution, the interactions that would reduce the van't Hoff factor are between $+1$ cations and -1 anions. In the copper(II) sulfate solution, the interactions are between $+2$ cations and -2 anions, because they have greater charges. These interactions are much stronger.

252. (B) When 1 mole of ammonium sulfate dissociates, it forms 3 moles of ions. Therefore, the van't Hoff factor is 3.

253. (B) Because all three are covalently bonded, they should have enough vapor pressure for fractional distillation to be effective, all other factors being equal, the first compound to be isolated is the one with the lowest molar mass. AsF_3 has a molar mass of 132 g/mol, PF_3 is 88 g/mol, and SbF_3 is 179 g/mol.

254. (A) There are a number of precipitates formed, so you should look for a solution that contains an ion that forms insoluble substances with the solutions provided. This is true only of $AgNO_3$. It forms white $AgCl$, yellow AgI, and black AgS.

255. (C) The transition metal carbonates will react with an acid to form carbon dioxide and water. Answer C is nitric acid, a strong acid.

Chapter 7: Kinetics

256. (B) When the temperature decreases, the rate of reaction decreases, and the rate constant also decreases. Answers A and D do not make sense; from your personal experience, cooling a reaction (even the bacterial growth in foods) will slow down the reaction. Answer C is incorrect because the rate constant does decrease with a decrease in temperature.

257. (B) Comparing Experiments 1 and 2, [B] is constant, [A] doubles, and the rate doubles. The reaction is first order in [A]. Comparing Experiments 2 and 3, the [A] is constant, the [B] doubles, and the rate doubles. The reaction is first order also in [B]. The rate law is Rate = k[A][B].

258. (A) For a reaction to be slower, the activation energy for the reaction must be higher. It is the magnitude of the activation energy that determines the speed of reaction.

259. (B) The rate constant is dependent on the activation energy: the larger the activation energy, the slower the reaction. The temperature also has an effect: increasing the temperature normally increases the rate constant. A catalyst lowers the activation energy and thus affects the rate constant. However, the enthalpy change of the reaction is not related to the rate constant.

260. (A) The acids in answers B, C, and D are all weak acids, and the concentration of the hydrogen ion will be well below 0.1 M. HCl, however, is a strong acid, so the hydrogen ion concentration will be 0.1 M. This is the highest concentration of hydrogen ion, so the HCl solution will cause the release of carbon dioxide the fastest.

261. (D) NO is consumed and then re-formed during the reaction, so even a small amount can react with a significant amount of ozone, making answer A false. [NO] is not involved in the rate-determining step, making answer B false. [NO] does affect the overall energy change, making answer C false.

262. (D) Changing concentrations of the reactants or products does not change the value of the rate constant, but lowering the activation energy will.

263. (A) Answer B would be true for most mechanisms; answer C would be true if the rate-determining step involved Cl_2; and the temperature dependence is not related to a particular mechanism. The detection of the ·OH supports the proposed mechanism.

264. (D) The reaction involves breaking the $H-O$ bond in water, so changing the bond strength would affect the rate. Since the reaction is promoted by light, changing the intensity would have a large effect on the rate. Free radicals are formed during the reaction, so adding a substance that removes free radicals would have a large effect. Adding a nonreacting gas would have a minor effect on the rate, since it would affect the number of effective collisions.

265. (C) In profile I, the products are at a lower energy than the reactants (exothermic), and in profile II, the products are of higher energy than the reactants (endothermic).

266. (A) In all but answer A, the products are of lower energy than the reactants, indicating an exothermic reaction. In answer A, the products are of higher energy, an endothermic reaction, and energy will have to be constantly supplied to make the reaction proceed.

267. (A) The rate-determining step should have the highest energy in the energy profile. If the rate-determining step is the first step, then the highest peak should be first. This is what appears in answer A.

268. (D) If [A] is doubled, the rate will increase fourfold ($2^2 = 4$); if [B] is doubled, the rate will double ($2^1 = 2$). If both happen at the same time, it will be an eightfold increase (4×2).

269. (A) Comparing Experiments 1 and 2, the $[I^-]$ was doubled (other concentrations remaining constant) and the rate doubled. Therefore, the reaction is first order with respect to the iodide ion. Comparing Experiments 3 and 1, the $[ClO^-]$ was doubled and the rate doubled; therefore, the reaction is first order in hypochlorite. Comparing Experiments 1 and 5, the $[OH^-]$ decreases by one-half and the rate doubles, meaning that the reaction is -1 with respect to the hydroxide ion.

270. (B) Increasing the hydroxide ion concentration causes the rate to decrease. Substances that do this are inhibitors.

271. (D) Rate = $k[CH_3CHO]^2$; substituting gives 1.0×10^{-2} $M\,s^{-1} = k[0.20]^2$; $k = (1.0 \times 10^{-2}\ M^{-1}\,s^{-1})\ /\ (0.04\ M)^2$. K = $0.25\ M^{-1}\,s^{-1}$.

272. (B) For the endothermic reaction, you would need to put in 45 kJ/mol to get to the energy level of the reactants and then an additional 85 kJ/mol to get to the top of the activation energy hill. This will be a total of 130 kJ/mol.

273. (A) Since only one oxygen molecule is produced for every two peroxide molecules that react, the rate of appearance of oxygen should be half the rate of disappearance of peroxide, or $(2.0 \times 10^{-1}\ M/s)\ /\ 2 = 1.0 \times 10^{-1}\ M/s$.

274. (B) Looking at the energy profile, you can see that the activation energy for the forward reaction is much greater than for the reverse reaction; therefore, the reverse reaction should be faster than the forward reaction.

275. (A) The reaction rate is changed by changes in temperature, which eliminates answer D, and by changes in concentration and pressure of reactants, which eliminates answer B. Decreasing the calcium carbonate particle size will increase the surface area and increase the reaction rate. Changing the concentration of a product does not change the rate of reaction. (This is a kinetics problem, not an equilibrium one.)

276. (D) The OH radical is formed in the rate-determining step, but it is consumed in the next step. This makes it a reaction intermediate.

277. (C) A catalyst increases the reaction rate by lowering the activation energy of a reaction. This corresponds to the height of the peak in the energy profile.

278. (D) The activation energy of the forward reaction is less than the activation energy of the reverse reaction; therefore, the forward reaction is faster than the reverse reaction. However, there is insufficient thermodynamic data to make a prediction (the entropy change is unknown).

279. (B) The activated complex appears at the top of the activation energy "hill" at position B.

280. (D) There is a general rule that says for a 10°C increase in temperature, the reaction rate doubles. Conversely, if the temperature were reduced 10°C, the rate would decrease by half. Answers B and C can immediately be eliminated, since most biological reactions will slow down if the temperature is decreased.

281. (D) The reaction with the fastest rate will be the one with the smallest activation energy. There is no correlation between ΔH and reaction rate. Be very careful of MCAT questions that mix thermodynamics with kinetics. Very few thermodynamic arguments give correct kinetics answers, and vice versa.

282. (D) Br is formed in one step and consumed in the next step. That makes it an intermediate.

283. (A) Comparing Experiments 1 and 3, the oxygen concentration is constant, the NO concentration doubles, and the rate increases fourfold. That means the rate is dependent on $[NO]^2$. Comparing Experiments 3 and 2, where the NO concentration is constant, the $[O_2]$ doubles and the rate doubles. The reaction is first order in $[O_2]$. Therefore, the rate law is Rate = $k[NO]^2[O_2]$.

284. (B) The rate-determining step in a reaction is the slowest step in the reaction mechanism.

285. (A) Comparing Experiments 1 and 2, where [B] and [C] are constant, the [A] doubles and the rate doubles. The reaction is first order in [A]. Comparing Experiments 2 and 3, where the [A] and [C] are constant, the [B] doubles and the rate doubles. The reaction is first order in [B]. Comparing Experiments 1 and 4, where the [A] and [B] are constant, the [C] triples and the reaction increases ninefold. The reaction is second order in [C]. Therefore, the rate law is Rate = $k[A][B][C]^2$.

286. (D) Answers A and C are the same: a catalyst lowers the activation energy and affects the rate constant. Since both cannot be correct, this eliminates both. Increasing the temperature also affects the rate constant. However, changing a reactant concentration does not change the rate constant.

287. (B) The rate of reaction is dependent on the activation energy: the larger the activation energy, the slower the reaction. The temperature also has an effect: increasing the temperature normally increases the rate of reaction. A catalyst lowers the activation energy and thus affects the reaction rate. However, the enthalpy change of the reaction is not related to the rate of reaction.

288. (A) One way to express the rate of reaction is to express the change in reactant concentration with time, eliminating answers C and D; this form has a minus sign and is multiplied by 1/a, where a is the coefficient in the balanced equation, eliminating answer B.

289. (C) There is a change in the oxidation state of chlorine, from 0 to -1; the reaction will be faster at elevated temperatures; and both a strong acid (HCl) and a weak acid (HClO) are formed. The equilibrium constant is not affected by changes in concentrations/pressures. (Remember that equilibrium studies are *separate* from kinetic studies.)

290. (B) The rate law should involve the rate-determining (slowest) step. This eliminates answer A. Answer D is an equilibrium constant, so it is eliminated. Answer C is mixing reactants and products. Answer B has the [H$_2$O] from the slow step and represents the [Cl·] as a function of the Cl$_2$ by using the one-half exponent.

291. (B) Answer A, the equilibrium answer, is related to thermodynamics, but this is a kinetics question. (This is like comparing apples and oranges and is a common "trick" on the MCAT.) Answers C and D also have nothing to do with rates of reaction. If the activation energy for reaction 2 is greater than the activation energy for reaction 1, the rate of 2 will be slower than that of 1.

292. (A) The correct energy profile would be one in which there was a small peak (first fast reaction), then a much higher peak (second slow reaction), and finally a smaller peak for the third fast reaction.

293. (A) Two moles of permanganate will disappear while 5 moles of iodine appears. The rate of appearance of iodine, then, should be 5/2 that of the disappearance of permanganate. Therefore, $(5/2) (2.0 \times 10^{-3} \ M/s) = 5.0 \times 10^{-3} \ M/s$.

294. (C) Since the drug is following first-order kinetics, in 4 hours (1 $t_{1/2}$), the level would be down to 50%; in 8 hours (2 $t_{1/2}$), 25%; in 12 hours (3 $t_{1/2}$), it would be at 12.5%; and at 16 hours (4 $t_{1/2}$), it would be at 6.25%. 12 hours would be the closest to 10%.

Chapter 8: Equilibrium

295. (D) Notice that answers B and D are exactly opposite, so one of these must be the true statement. Look at the K_{sp} values of the two sparingly soluble salts; AgOCN has a larger K_{sp} than AgCN, and therefore, it is more soluble. So answer D is the correct answer. Answer C is false because the presence of a common ion reduces solubility, not increases it. In addition, the ΔG of the AgOCN should be more positive than for the AgCN, since its solubility is much greater.

296. (A) The form of the K_{sp} would be $K_{sp} = [Ag^+][C_2H_3O_2^-]$. Precipitation will occur at the point at which the solubility product constant is reached. You have the K_{sp} value and the $[Ag^+]$, so you can use the solubility product constant and solve for the acetate ion concentration: $4 \times 10^{-3} = [1 \times 10^{-3}][C_2H_3O_2^-]$. Therefore, $[C_2H_3O_2^-] = 4 \times 10^{-3} / 1 \times 10^{-3} = 4 \ M$.

297. (C) The equation for the dissolving of the calcium hydroxide is $Ca(OH)_2 \leftrightarrows Ca^{2+} + 2 \ OH^-$, so the K_{sp} expression would be $K_{sp} = [Ca^{2+}][OH^-]^2$.

298. (D) The equilibrium equation is $CaCO_3 \leftrightarrows Ca^{2+} + CO_3^{2-}$. Adding sodium carbonate, an ion common to the equilibrium, forces the equilibrium back to the left, increasing the amount of solid calcium carbonate.

299. (C) Increasing the pressure would cause the equilibrium to shift to the side with the fewest number of moles of gas, the left, decreasing the amounts of CO and H_2 and increasing the amounts of CH_4 and H_2O.

300. (D) Answer A is true; adding a catalyst does not change the value of the equilibrium constant. Answer B is true; adding additional reactant or product to an equilibrium affects the equilibrium amounts and is important. Answer C is true; adding a solid reactant will not affect the equilibrium. Answer D is false; adding a catalyst does not affect the value of an equilibrium constant.

301. (B) The equilibrium constant would have the form of $P_{SO_3}^2 / (P_{SO_2})^2(P_{O_2})$. Substituting the pressures (and ignoring units) gives $(10.0)^2 / (10.0)^2(5) = 0.2$. Notice that the $(10.0)^2$ in both the numerator and denominator would cancel—no need to do that calculation.

302. (A) A catalyst affects the speed at which a reaction occurs, not the final result. The reaction will reach equilibrium faster, but the equilibrium constant will be the same.

303. (B) The oxalate ion is the conjugate base of oxalic acid. This basic ion would be stable only in a basic pH. Therefore, the correct answer is in the high (basic) blood pH. Also note that A and C are the same, which eliminates both answers.

304. (D) The equilibrium is established with I-131, and then additional I-127 is added. Even though it is a different isotope of iodine, chemically it will react the same and force the equilibrium to the left; however, some of the $AgI(s)$ now contains ^{127}I, which lowers the fraction of ^{131}I.

305. (A) By the solubility rules, the only one of the four choices that is insoluble is lead sulfate.

306. (C) The double arrow indicates an equilibrium state.

307. (A) Increasing the temperature favors the endothermic reaction; therefore, more H_3O^+ is formed, so the pH should slightly decrease.

308. (C) The addition of sulfuric acid will increase the $[H^+]$, which will cause both equilibriums to shift to the left (eliminating answer D as false), lowering the pH (eliminating answer A as false), and since both equilibriums shift to the left, the amount of undissociated H_2S will increase (eliminating answer B as false). There is no reaction to produce H_2 possible with these reactions.

309. (B) If the pressure is reduced, the equilibrium will shift to the side with the greater number of moles of gas to increase the pressure. In this specific equilibrium, there are 2 moles of gas on the left and only 1 mole of gas on the right. Therefore, the reaction will shift to the left.

310. (D) Increasing the temperature favors the endothermic side of the equilibrium, in this case the left side. Increasing the temperature will cause the equilibrium to shift to the left.

311. (A) Air, with its normal concentration of carbon dioxide and oxygen, is taken into the lungs. There, carbon dioxide from the cells is exchanged for oxygen. The exhaled gas is thus lower in oxygen (since some went to the cells) and higher in carbon dioxide (since more was added from the cells).

312. (A) Reducing the volume really is a pressure effect. If the volume is reduced, the pressure increases. Answer C is incorrect, since it involves aqueous solutions. Answers A, B, and D involve gases. In the case of answer A, it is a change of 3 moles of gas; in answers B and D, it is a 1-mole change of gas. The greater effect will be in the 3-mole change, answer A.

313. (B) The reaction quotient has the same form as the equilibrium constant. Tripling a concentration that appears in the numerator will triple the reaction quotient.

314. (A) $HClO$ is not a strong acid, so this eliminates answers C and D. The dissociation of $HClO_3$, a strong acid, will reduce the concentration of $HClO_3$, and the equilibrium will shift to the right to replace it.

315. (B) Precipitation will occur if the reaction quotient exceeds the K_{sp}. With rounding, the $[Ca^{2+}]$ would be (10.0 mL $\times$ 0.01 M) / 30.0 mL = 3×10^{-3}, and the $[F^-]$ would be (20.0 mL $\times$ 0.01 M) / 30.0 mL = 7×10^{-3}. The reaction quotient would be $[Ca^{2+}][F^-]^2 = [3 \times 10^{-3}][7 \times 10^{-3}]^2 = 1.5 \times 10^{-7}$. The reaction quotient is less than the K_{sp}, so no precipitation will occur.

316. (C) This is a common ion problem. Answer D would not furnish a common ion (Cl^-), because the chlorines are covalently bonded. The other three answers will furnish chloride ions, but the calcium chloride will furnish twice as many chloride ions as the other two answers and will therefore decrease the solubility the most.

317. (A) Heating the mixture would facilitate the decomposition of the calcium carbonate; adding carbon dioxide would shift the equilibrium to the left; and decreasing the volume would increase the pressure, shifting the equilibrium to the left. Adding CaO, a solid, would have very little effect on the equilibrium.

318. (D) Changing the temperature affects the equilibrium, as does changing the amounts of the gases. Adding a catalyst, however, affects both the forward and reverse reactions equally, resulting in no net change.

319. (B) Changing the temperature affects the equilibrium, as does changing the amounts of the gases, but since both sides have an equal number of moles, changing the pressure will have no effect.

320. (B) The equilibrium constant has the form of $K_c = [NO]^2 / [N_2][O_2]$. Substituting the concentrations gives $K_c = [0.20]^2 / [0.10][0.20] = [0.20]/[0.10] = 2.0$.

321. (B) This is a common ion problem. NaCN is added, and the $[CN^-]$ increases, shifting the equilibrium to the left and therefore decreasing the solubility of the AgCN, causing more of it to precipitate. The equilibrium constant does not change.

322. (C) The form of the solubility product constant would be $K_{sp} = [Ag^+][CN^-]$. If 1.0×10^{-8} mol/L of AgCN dissolves, the $[Ag^+] = [CN^-] = 1.0 \times 10^{-8}$. Substituting into the K_{sp} expression gives $K_{sp} = [1.0 \times 10^{-8}][1.0 \times 10^{-8}] = 1.0 \times 10^{-16}$.

323. (A) This is a common ion problem. Calcium sulfate is the sparingly soluble salt, so that adding it to a sodium sulfate solution increases the sulfate ion concentration and shifts the equilibrium back to the left, decreasing the solubility of the calcium sulfate.

324. (A) The equilibrium constant expression is a fraction with the product of the right-hand species, each concentration raised to the power of the coefficient in the balanced chemical equation, divided by the product of the left-hand species, each concentration raised to the power of the coefficient in the balanced chemical equation. Therefore, the equilibrium expression would be $K_c = [NO_2]^2 / [NO]^2[O_2]$.

325. (B) Adding CO would force the equilibrium back to the left, decreasing H_2 production. Removing CH_4 would also shift the equilibrium back to the left to replace the methane that was removed, decreasing H_2 production. Increasing the pressure would shift the equilibrium to the side with the fewest number of moles of gas to decrease the pressure: the equilibrium shifts to the left, decreasing H_2 production. Removing H_2 would shift the equilibrium to the right to replace the hydrogen gas, thus increasing H_2 production.

326. (A) Catalysts increase the rates of reaction, allowing the system to reach equilibrium faster, but do not affect the equilibrium concentrations of the reactants and products.

327. (A) The negative ΔH means that the forward reaction is exothermic. Answer B is false, since increasing the temperature will shift the reaction to the endothermic side, the left in this case, reducing the ammonia concentration. Answer C is false because changing the temperature will disturb the equilibrium. Answer D is false because increasing the temperature will shift the reaction to the left when there are more moles of gas and thus will increase the pressure. Answer A is correct; the kinetic energy of the molecules will increase at a higher temperature.

328. (B) You know initially that the sulfur dioxide and oxygen are in a 2:1 ratio and that the total pressure is 30 atm. That means that the partial pressure of sulfur dioxide is 20 atm and oxygen's is 10 atm. At equilibrium, $P_{SO_2} + P_{O_2} + P_{SO_3} = 25$ atm. If you let x = pressure of SO_3, then the partial pressure of SO_2 is $(20 - x)$ and the partial pressure of O_2 is $(10 - 0.5x)$. (1:2 stoichiometry between O_2 and SO_3.) Putting these values into the previous equation, $P_{SO_2} + P_{O_2} + P_{SO_3} = 25$ atm, gives $(20 - x)$ atm $+ (10 - 0.5x)$ atm $+ x$ atm $= 25$ atm. Solving for x gives $-0.5\, x = -5$, or $x = 10.0$ atm.

329. (D) The solubility of the calcium fluoride is changed by the presence of a common ion, shifting the equilibrium back to the left and reducing the solubility. The only solution that has a common ion is the calcium nitrate.

330. (C) The formula of cupric arsenate is $Cu_3(AsO_4)_2$, and the K_{sp} expression is $K_{sp} = [Cu^{2+}]^3[AsO_4^{3-}]^2$, since when the salt dissolves, you get 3 cupric ions and 2 arsenate ions. If you let x = moles per liter of the salt that dissolves, then $[Cu^{2+}] = 3x$ and $[AsO_4^{3-}] = 2x$. Substituting into the K_{sp} expression gives $K_{sp} = (3x)^3(2x)^2 = 10^{-37}$. To keep from doing a lot of math, you can approximate $(27x^3)(4x^2) \neq 100x^5$, giving a K_{sp} expression of $10^{-37} = 100x^5$, or $10^{-35} = x^5$. Therefore, $x = 10^{-7}$, $[Cu^{2+}] = 3x = 3 \times 10^{-7}\ M$.

331. (B) The equilibrium is established with I-131, and then additional I-127 is added. Even though it is a different isotope of iodine, chemically it will react the same and force the equilibrium to the left, reducing the concentration of I-131.

332. (D) Calcium carbonate, $CaCO_3$, is an insoluble salt, so the concentration of the solution would be very low. The <0.6 molar would be the most logical choice.

Chapter 9: Acids and Bases

333. (D) Metal oxides in general, when dissolved in water, tend to yield basic solutions due to the formation of hydroxides. However, oxides of metals that have a high oxidation state tend to form acidic solutions. Mn_2O_7 is in the highest oxidation state for manganese.

334. (A) The reaction is $HPO_4^{2-} \leftrightarrows H^+$ and PO_4^{3-}. Therefore, the phosphate ion is the correct choice.

335. (B) A buffer has both a weak acid and a weak base component. Answers A, C, and D do not provide a weak base to be with the weak nitrous acid. However, if you add some hydroxide ion, it will react with the nitrous acid to form its conjugate base, the nitrite ion. Now you have both a weak acid and its conjugate base, a buffer.

336. (A) The hydrogen halides would have the formula HX, where X is any halogen (F, Cl, Br, I, or At). When the hydrogen halide dissociates in water, the anion of the halogen results. F^- is the only halide anion in the answers.

337. (B) No precipitate will form, since no cation that will form an insoluble salt is added. You can form a gas from a carbonate ion solution by adding an acid, forming H_2CO_3, which decomposes to carbon dioxide and water.

338. (A) CCl_3COOH would be the strongest, since it has three chlorines attached to the noncarbonyl carbon. These chlorines are extremely electronegative, pulling electron density from the oxygen atoms and further weakening the $O-H$ bond, making it a stronger acid.

339. (B) HI is a strong acid, so the $[H^+] = 10^{-3}$ would give a pH of 3.

340. (A) Every pH unit difference amounts to a power of 10 difference. There is a difference of 6 pH units, so that would be a difference in hydrogen ion concentration of 10^6.

341. (B) Hydrochloric acid is a strong acid. There has been a tenfold dilution of the volume (10 mL to 100 mL). This will reduce the hydrogen ion concentration by 1 power of 10 and therefore 1 pH unit. The hydrogen ion concentration has been reduced, so the pH should increase from a pH of 1 to 2.

342. (A) There should be no appreciable reaction with the SiO_2 (the principal component of sand). Calcium carbonate might react, but it would result in a basic solution, since the carbonate ion is a weak base. PbS is insoluble, so no appreciable reaction would be expected. Sulfur dioxide reacts with water to form an acidic solution of sulfurous acid, which may undergo further oxidation to sulfuric acid.

343. (B) For a solution to buffer against both an acid and a base, there must be both an acid component and a base component. Sodium carbonate will react only with an acid; the ammonium chloride will react only with a base; and calcium chloride will react with neither an acid nor a base. The sodium bicarbonate is amphoteric (amphiprotic) and will react with an acid or a base. It reacts with an acid to form carbonic acid (which decomposes to carbon dioxide and water) and with a base to form the carbonate anion.

344. (A) This is a slightly basic pH, so answer B is true. The hydrogen ion concentration is less than the hydroxide ion concentration: pH + pOH = 14, so pOH = 14 − pH, or pOH = 14 − 7.4 = 6.6. So answer C is true. Organic acids would be more soluble in this basic solution than in water, so answer D is true. Calcium carbonate, however, would not be soluble in a basic solution, since the carbonate ion is also a base.

345. (D) Both sulfonic acids and carboxylic acids have acid-base properties that can be used as an indicator. An amine group is basic, so it may also be used in an indicator. An alcohol has no appreciable acid-base properties and thus cannot be used as an acid-base indicator.

346. (A) The reaction is $CH_3NH_3^+ \leftrightarrows H^+ + CH_3NH_2$, and the $K_a = [H^+][CH_3NH_2] / [CH_3NH_3^+]$. If you let x = number of moles per liter of $CH_3NH_3^+$ that dissociate, then $x = [H^+] = [CH_3NH_2]$, and $[CH_3NH_3^+] = 1 - x$. If x is small compared with 1, then $1 - x \neq 1$, and substituting into the K_a expression gives $K_a = [H^+][CH_3NH_2] / [CH_3NH_3^+]$. Substituting gives $2 \times 10^{-12} = [x]^2 / [1]$. Therefore, $x = 1 \times 10^{-6} = [H^+]$. The pH is 6.

347. (C) Answers B and D say the same thing, so neither can be correct. You will be reducing the hydrogen ions but replacing them with potassium ions, so the number of ions remains the same and the conductivity remains constant. Because you are adding a base that reduces the number of hydrogen ions, the pH should increase, not decrease.

348. (C) Nitric acid is a strong acid and had ionized completely, so the $[H^+]$ = 0.01 M. The sodium hydroxide reacts with 99% of the initial H^+, but 1% (1/100, or 0.01) of the original hydrogen ion solution remains, so $0.01 \times 0.1 M = 1 \times 10^{-3}$ is the concentration of the remaining H^+. The pH is therefore 3.

349. (A) The buffer capacity is the amount of either acid or base that can be neutralized without appreciably changing the pH. The more acid and base present, the higher the buffer capacity. Therefore, it is advantageous to maximize the concentrations of the conjugate acid-base pair.

350. (A) Answer B is simply another way of writing phosphoric acid, a polyprotic acid. Ammonia is a monoprotic base, and answer D is sulfuric acid, a polyprotic acid. CaO, when dissolved in water, yields calcium hydroxide, $Ca(OH)_2$, which is a polyprotic base.

351. (D) The production of water from hydrogen and hydroxide ions is exothermic; therefore, the reverse reaction, production of the ions from water, is endothermic. It is on this endothermic reaction that the K_w is based. Heating a reaction favors the endothermic reactions, so the concentration of hydrogen and hydroxide ions would increase, thereby increasing the value of the K_w.

352. (A) All of the halides except F are conjugate bases of strong acid and will not hydrolyze. However, the fluoride ion is the conjugate base of the weak acid HF and therefore will react with water to produce hydroxide ions, which will make the solution slightly basic: $F^- + H_2O \leftrightarrows HF + OH^-$.

353. (C) There are eight orders of magnitude difference in pH between the two solutions, which is a 10^8 difference in $[H^+]$.

354. (C) Since a strong acid ionizes almost completely, there is a high concentration of products and a very low concentration of reactants, so that the $K_a > 1$.

355. (B) In a polyprotic system, each successive ionization is much smaller than the previous. This eliminates answers A, C, and D. The $[H^+]$ should be the largest, since it is the product of all three ionizations.

356. (B) Since the solution formed was acidic, the $K_a > K_b$.

357. (A) From the table, it is apparent that the smaller the bond energy, the larger the K_a (the stronger the acid). The O, S, Se, and Te are in the same family in order from top to bottom on the periodic table. The electronegativities are decreasing in this order. Therefore, the acidity (strength) increases as bond energy and electronegativity decrease.

358. (C) All of the hydrogen halides are strong acids with the exception of HF, which is a weak acid. This eliminates answers A, B, and D. In general, the acidity of the hydrogen halides is inversely proportional to the electronegativities of the element. Electronegativities decrease down a family, so acidity should increase.

359. (A) From the K_a values, you can see that HClO is the weaker acid, so the ClO^- is the stronger base. Therefore, it will protonate (accept an H^+) to a greater extent.

360. (D) Answers A, B, and C are bases and would react with the acids, so you could not differentiate between the two. $HC_2H_3O_2(l)$ is the only acid, and it could act as a solvent.

361. (B) Since amino acids can act as both acids and bases, they are considered amphoteric (amphiprotic).

362. (C) The reaction would be $NaOH(aq) + HNO_3(aq) \rightarrow H_2O(l) + NaNO_3(aq)$. Since it is the reaction between a strong base and a strong acid, it essentially goes to completion. The amounts and concentrations were equal, so it is a complete neutralization, with a resulting pH of 7.

363. (A) Looking at the reaction, it is obvious that sulfuric acid is donating a hydrogen ion to nitric acid. That makes the nitric acid a base, a hydrogen ion acceptor.

364. (B) The reaction is HOCN $\leftrightarrows$ H$^+$ + OCN$^-$, and K$_a$ = [H$^+$][OCN$^-$] / [HOCN]. If you let x = number of moles per liter of HOCN that dissociate, then x = [H$^+$] = [OCN$^-$], and [HOCN] = 1 − x. If x is small compared with 1, then 1 − x ≠ 1, and substituting into the K$_a$ expression gives 3.5 × 10^{-4} = [x][x]/[1]. [OCN$^-$] = x = 1.9 × 10^{-2}. (Simplifying by estimating gives x^2 = 10^{-4}, which, by inspection, gives x = 10^{-2}.)

365. (A) The equilibrium is HOCN $\leftrightarrows$ H$^+$ + OCN$^-$. Adding additional cyanate anion will cause the reaction to shift to the left, reducing the [H$^+$] and increasing the pH. (Remember that increasing the [H$^+$] lowers the pH, and vice versa.)

366. (D) According to the Henderson-Hasselbalch equation, pH = pK$_a$ + log([base]/ [acid]). In this case, the concentrations of the acid and base are equal, so the pH will equal the pK$_a$ (2 × 10^{-6}). If the pK$_a$ was 1 × 10^{-6}, then the pK$_a$ would be 6 and so would the pH. Since the K$_a$ is a little greater, then the pK$_a$ and the pH will be between 5 and 6.

367. (C) HNO$_3$ is a strong acid, so the [H$^+$] = 0.010, or 1 × 10^{-2}. The pH would then be 2.0.

368. (C) The hydroxide ion is a strong base. All the rest are weak bases.

369. (D) C$_2$H$_5$OH is a nonelectrolyte—no base formation. H$_3$PO$_4$ is an acid; it will form an acidic solution. NaCl is a strong electrolyte—just a neutral salt solution formed. Na$_2$CO$_3$, when it dissolves, will release carbonate ions into the solution. Carbonate ions are weakly basic and will react with the water to form the bicarbonate ion and hydroxide, resulting in a basic solution.

370. (C) Conjugate acid-base pairs differ in a single H$^+$. All of the answers except answer C fulfill this requirement. In answer C, there is a difference of a hydrogen ion, but the anion portion has changed from nitrate to nitrite.

371. (C) The pK$_a$ is the −log K$_a$. The second equilibrium is very small and can be ignored, so only K$_{a1}$ is considered. The second equilibrium does not involve H$_2$S, so it is irrelevant to the question. Therefore, −log(K$_{a1}$) = −log(1 × 10^{-7}) = 7.

372. (C) Ammonium chloride, when dissolved, releases the weak acid, the ammonium ion. This would increase the hydrogen ion concentration (and lower the hydroxide concentration). Neither ethyl alcohol nor sodium chloride would affect the pH of the water. The potassium phosphate, when dissolved, releases the phosphate ion, a weak base. This would react with water to form hydroxide ions, increasing the hydroxide ion concentration.

373. (B) For the titration of a weak acid with a strong base, a weak base will be the predominant species at the equivalence point. An indicator should therefore change color at a slightly basic pH. Thymol blue is the only indicator to change color in a basic range.

374. (C) According to the Henderson-Hasselbalch equation, $pH = pK_a + \log([base]/[acid])$. The $pK_a \neq 12$ (assuming $K_a \neq 1 \times 10^{-12}$), and the $([base]/[acid]) = 1$. Therefore, $pH = 12 + 0 = 12$.

375. (A) A buffer is composed of a mixture of a weak acid and a weak base. (Commonly, a conjugate acid-base pair differ by a single H^+.) HCl is a strong acid; CaO and $Ca(OH)_2$ are both basic; and $C_2H_5OH_2^+Cl^-$ and C_2H_5OH are not acid-base pairs. However, HCO_3^- and CO_3^{2-} are a conjugate acid-base pair and thus will act as a buffer.

376. (A) Mixing an equal volume of the solutions cuts their concentrations in half, making each solution now $0.25 M$. The nitric acid, a strong acid, will react with the ammonia, a weak base, and protonate all of the ammonia to ammonium ions. Left in solution will be ammonium ions and nitrate ions with a concentration of $0.25 M$ each. The pH will be in the acidic range due to the presence of the weak acid, the ammonium ion.

377. (B) The entire discussion of the nitrite ion is meant to distract you as the real exam might. The problem stated that the iron was originally in the $+2$ state before undergoing oxidation.

378. (A) The strongest base would have the smallest K_a of its conjugate acid.

Chapter 10: Thermodynamics

379. (A) During the reaction, you are going from 2 moles of gas to 1, thereby decreasing the entropy of the system.

380. (D) This is a thermodynamic problem; however, answers A, B, and C refer to kinetics, not thermodynamics. Changing the temperature will affect thermodynamic properties such as $\Delta G°$. It is important to keep the two, kinetics and thermodynamics, separate.

381. (C) The best indicator of whether a reaction is spontaneous is the change in the Gibbs free energy, so you can eliminate the first two answers. If a reaction is at equilibrium, then $\Delta G = 0$; this eliminates answer D. If a reaction is spontaneous, its ΔG is negative, and in a nonspontaneous reaction it is positive.

382. (B) Gases, because of their freedom of motion, have greater entropies than solids. Both ΔH_f° and ΔG_f° for any element in its standard state is zero. The standard state for gases (STP) is not the same as the thermodynamics standard state. Statement B is true only for a crystalline substance and not for other substances, making the statement false.

383. (A) The $\Delta H_{rxn} = \Delta H_f^\circ$ of the products $- \Delta H_f^\circ$ of the reactants. Both reactants are in their standard state, so all you have to consider is the products. Therefore, $\Delta H_{rxn} = (2 \text{ mol})(-270 \text{ kJ/mol}) = -540 \text{ kJ/mol}$.

384. (C) For a spontaneous process, the enthalpy change is negative and the combustion process has a large positive entropy change. Together, the enthalpy and entropy yield a very negative ΔG. A large negative ΔG gives rise to a large positive K (there are no negative Ks) through $\Delta G = -RT \ln K$.

385. (D) An adiabatic process is one in which there is no net heat exchange between the surroundings and the system.

386. (D) The foam coffee cup will not withstand very high temperatures and cannot be sealed to effectively handle high or low pressures. Therefore, A, B, and C are all true, and the correct answer is D, all of the above.

387. (A) Substituting a weak acid for a strong acid would mean that fewer particles would be consumed because of the weak base equilibrium that would be established. There would be less of an enthalpy change and therefore less of a temperature change.

388. (D) During the reaction, a mole of gas is lost (converted to a solid); therefore, the entropy should decrease.

389. (C) You are adding a solid, $SnCl_2(s)$, and solids, as long as some are present, do not affect the equilibrium.

390. (B) With respect to the enthalpy change, a spontaneous process has a negative sign; with respect to the entropy change, a spontaneous process has a positive sign. With those two considerations, a process is always spontaneous if the enthalpy change is negative and the entropy change is positive.

391. (D) The Gibbs free energy, which is negative for a spontaneous reaction, has the form of $\Delta G = \Delta H - T\Delta S$. If both the enthalpy and entropy changes were positive, then making T sufficiently large would give an overall negative sign to the Gibbs free energy change. Therefore, the reaction could become spontaneous. In all the other choices, increasing the temperature does not lead to a negative ΔG.

392. (C) The change in the internal energy (ΔE) is equal to the heat transferred (q) plus any work done (w): $\Delta E = q + w$. Since heat is being removed, it has a negative sign, and since work is being done on the system, w has a positive sign. Therefore, $\Delta E = -375 \text{ J} + 1{,}475 \text{ J} = 1{,}100 \text{ J}$.

393. (B) Spontaneity is related to both the entropy and enthalpy changes in a system. A negative entropy change can be offset by a large change in the enthalpy.

394. (A) When methanol evaporates, it must absorb heat, and therefore, the enthalpy change is positive; as the methanol goes from the liquid to the gaseous state, the entropy increases, resulting in a positive sign for the entropy change.

395. (C) When carbon tetrachloride vaporizes, the intermolecular forces must be broken. Carbon tetrachloride is nonpolar, and because of this, the predominant intermolecular force is London dispersion forces.

396. (B) Since ΔG is positive, the forward reaction is nonspontaneous, and so the equilibrium constant, K, will be less than 1.

397. (A) 4.0 g of Ca is 0.10 mole. (Ca = 40 g/mol.) From the reaction, you see that there is a 1:1 stoichiometry between calcium and hydrogen gas, so 0.10 moles of hydrogen gas will be formed. Since the gas is at STP, you know that 1 mol = 22.4 L, so 0.10 mol = 2.24 L. Substituting this volume and pressure into the equation 1 J = 100 L·atm, you get (100)(2.24 L)(1 atm) = 224 J. Since work is being done by the system, the value is negative.

398. (B) The overall enthalpy change would be the difference in the enthalpy of the reactants and products, which is labeled II on the figure.

399. (D) The $\Delta H°$ for the reaction is ($\Delta H_f°$ of products) $-$ ($\Delta H_f°$ of reactants) = [1 mol (-233 kJ/mol)] $-$ [1 mol (-110 kJ/mol) + 1 mol (0 kJ/mol)] = -113 kJ.

400. (B) $\Delta G = \Delta H - T\Delta S$. To ensure that ΔG is positive, ΔH must be positive, and ΔS must be negative. If ΔS is negative, then ($-T\Delta S$) will be positive, making the $-\Delta G$ positive.

401. (A) Heat (q) is the heat transferred and is related to the internal energy of the system (ΔE) and the work done (w) by the equation $\Delta E = q + w$. This eliminates answers C and D. (Only answer B or C needs to be wrong in order for answer D to be wrong.) Overall, heat is absorbed in the process, making q positive.

402. (A) Answer A should have the greatest increase in entropy because there is a net increase of 6 moles of gas. That greatly increases the entropy of the system. In answers C and D, you are going from gases to gases (not a large increase in entropy), and in answer B, you are going from a liquid to gases (not as large an increase as going from a solid to a gas).

403. (A) Heating water causes an increase in the kinetic energy of the particles. This increase in kinetic energy will increase the entropy of the water molecules because of the increased freedom of motion of the particles. Answer B is incorrect because even at 75°C there is a great deal of hydrogen bonding. Answer C is incorrect because water vaporizes at 100°C, and answer D is incorrect because the viscosity actually decreases.

404. (A) The insulated foam coffee cup greatly reduces the loss of heat from the inside to the outside; that is, from the system to the surroundings. Thus, it is nearly adiabatic.

405. (A) Answer D is incorrect; both compounds are ionic. The formation of $Ba(g)$ yields a large increase in entropy, which makes answer C incorrect. Since the reaction is conducted at high temperatures, the reaction is most probably not spontaneous at room temperature, making answer B incorrect. In the reaction, silicon goes from 0 charge to a 4+ by losing electrons. This is oxidation.

406. (D) The burning of methane is exothermic, and therefore ΔH is negative, eliminating answers A and C. During the reaction, it goes from 5 moles of gas to 6, increasing the entropy (positive), eliminating answer B.

407. (C) In Experiment 3, the NaOH had to dissolve before the acid-base reaction takes place (as in Experiment 2). The dissolving of NaOH is exothermic, resulting in the release of a greater amount of heat and a higher temperature.

408. (B) Since there was no heat exchange with the surroundings, q = 0.

409. (A) The ΔG of a reaction at equilibrium equals 0.

410. (B) The enthalpy change is calculated by the (sum of enthalpies of products) − (sum of enthalpies of reactants) = [1 mol (−511 kJ/mol)] − [1 mol (−325 kJ/mol) + 1 mol (0 kJ/mol)] = −186 kJ.

411. (A) Given $\Delta S = -\Delta H/T$, $T = -\Delta H/-\Delta S$. Expressing the enthalpy in J/mol gives (−10,000 J/mol) / (−9.50 J/mol·K). The minus signs cancel, as do the J/mol, leaving T = 1,053 K. If you approximate the 9.5 as 10, you simplify the calculation and get 1,000 K.

Chapter 11: Electrochemistry

412. (B) $CuCl_2$ is a soluble ionic compound releasing ions upon dissolution. It is not easy to oxidize Cu^{2+}, so it is not likely to be a reducing agent. However, the copper cation can undergo reduction: $Cu^{2+} + 2e^- \rightarrow Cu$, which makes it the oxidizing agent, the species that facilitates oxidation.

413. (D) Increasing the amount of solid aluminum will not affect the cell potential, eliminating answers A and C. A positive cell potential indicates that the reaction is spontaneous as written. The reverse reaction will therefore be nonspontaneous.

414. (A) For pressure to have an effect on the cell potential, there must be gases involved; this is true only for reactions 1 and 3. Therefore, answer A is correct.

415. (A) Increasing the amount of solid silver will have no effect, because the concentrations of pure solids are not factors.

416. (D) In an electrolytic cell, you must supply an electrical current to cause a desired redox reaction to occur. Therefore, the cell reaction is nonspontaneous, eliminating answers A and C. By definition, oxidation occurs at the anode, eliminating answer B.

417. (A) The strongest reducing agent (species being oxidized) is the one that would have the most negative reduction half-cell potential. If you were to flip the reaction around to get the oxidation half-reaction, it would be the one with the greatest oxidation half-cell potential. It would, therefore, be Zn.

418. (B) Electrolysis of molten magnesium chloride gives magnesium metal and chlorine gas. The magnesium cations will be reduced (gain of electrons), and the chloride anions will undergo oxidation (loss of electrons). Oxidation is an anode process, and reduction is a cathode process.

419. (C) During the reaction, Ca is undergoing oxidation ($Ca \rightarrow Ca^{2+} + 2\,e^-$); that makes it the reducing agent, eliminating answer B. Oxidizing/reducing agents have to be reactant species, eliminating answer D. Cl is undergoing reduction ($Cl_2 + 2\,e^- \rightarrow 2\,Cl^-$), making it the oxidizing agent.

420. (A) The silver electrode undergoes the following reaction: $Ag \rightarrow Ag^+ + e^-$. This is oxidation and occurs at the anode, eliminating answers C and D. The other electrode must be the cathode and must have a negative charge to draw the silver ions to it in order to undergo reduction to silver metal on the object to be plated.

421. (C) The standard reduction potentials follow the same order within a group as the electronegativities. Fluorine is the most electronegative, and the electronegativities (and standard reduction potentials) decrease going down the group.

422. (B) Reduction occurs at the cathode. The list of half-reactions tells us that reduction will take place with a half-reaction that has a negative voltage. The first one that has a negative voltage is the water decomposition to hydrogen gas.

423. (D) The half-cell reactions are $2 H_2O + 2 e^- \rightarrow H_2 + 2 OH^-$ and $2 H_2O \rightarrow O_2 + 4 H^+ + 4 e^-$. The reaction using hydrogen cations is reduction and occurs at the cathode, and the reaction involving the water forming oxygen is oxidation and occurs at the anode.

424. (A) If the pH increases, the concentration of the hydrogen ion decreases and the cell potential decreases.

425. (A) 0.64 g of Cu is approximately 0.01 mole. (Cu = 63.55 g/mol.) Looking at the half-reaction, there is a 2:1 ratio of moles of electrons to moles of copper. Therefore, there will be 0.02 moles of electrons involved in the deposition. 0.02 moles of electrons = 0.02 faradays.

426. (B) The two balanced half-reactions are $Fe^{2+} \rightarrow Fe^{3+} + e^-$ and $6 e^- + 14 H^+ + Cr_2O_7^{2-} \rightarrow 2 Cr^{3+} + 7 H_2O$. When electron loss and gain between the two half-reactions are equalized (as they must be), the iron half-reaction will be multiplied by 6.

427. (A) To serve only as an oxidizing agent, the compound can never serve as a reducing agent. $KMnO_4$ contains Mn^{7+}, which is the highest oxidation state of manganese. Since there are no higher oxidation states for manganese than Mn^{7+}, it can never serve as a reducing agent. The other three compounds have manganese in an intermediate oxidation state; therefore, any of the three can undergo either oxidation or reduction. (They can behave as either an oxidizing or a reducing agent.)

428. (D) Hexavalent chromium would be chromium in the +6 oxidation state. Assigning oxidations to each of the elements in answers A, B, and C shows that the chromium is in the +6 oxidation state. By process of elimination, answer D has to be the correct answer (as chromium is trivalent).

429. (B) There is no electron exchange in the reaction, so it cannot be a redox reaction, and disproportionation is a type of redox reaction. Chromium is Cr^{6+} throughout. Therefore, the correct answer must be B.

430. (C) If a reaction is spontaneous, its Gibbs free energy change is negative, <0.

431. (B) Since you know that Mg metal is a product of the reaction, then the chlorine half-reaction is reversed and the sign changed on the voltage. Then you can add the reaction along with the half-cell potentials. You get -3.73 volts, which means you must apply a minimum of $+3.73$ volts.

432. (A) The overall cell reaction would be $SrCl_2 \rightarrow Sr(s) + Cl_2(g)$. Applying reaction stoichiometry (with rounding) gives:

$$\frac{8.8 \text{ g Sr}}{1} \times \frac{1 \text{ mol}}{88 \text{ g}} \times \frac{1 \text{ mol Cl}_2}{1 \text{ mol Sr}} = 0.10 \text{ mol Cl}_2$$

That gives approximately 0.1 moles of Sr and, with the 1:1 stoichiometry, 0.1 moles Cl_2.

433. (C) The reducing agent is the substance that is undergoing oxidation, loss of electrons. It is the Cl^- as seen in the oxidation half-reaction: $2 Cl^- \rightarrow Cl_2 + 2 e^-$.

434. (A) In the reaction, aluminum is being oxidized. Therefore, the voltage associated with aluminum is a +1.7 V. The overall cell potential is 1.3 V = 1.7 V + X; X, the half-cell potential associated with the indium, is −0.4 V.

435. (D) Sodium cannot be produced by electrolysis of an aqueous solution, because the sodium would react violently with water.

436. (C) Platinum is normally inert, resisting redox reaction and acid-base reactions. In order to create a cell, you must have a physical conducting surface on which the electrons can flow. It has to be something upon which one can hook a wire.

437. (A) Oxidation is the loss of electrons and/or numerical increase in oxidation number. In answer B, oxygen is going from the 0 to −2 oxidation state; in answer C, it remains −2; and in answer D, from the −½ to −1. In answer A, however, it goes from −1 to 0, an increase in oxidation number.

438. (C) In answer A, the electronegative fluorine atoms are bonded to the carbon; in answer B, the electronegative oxygen atoms are bonded to the sulfur; and in answer D, there are no 2 electronegative atoms in the structure. In answer C, however, the 2 electronegative atoms are bonded together, making it a good oxidizing agent.

439. (B) In the half-reaction generating the chromium(III) cation, $6 e^- + 14 H^+(aq) + Cr_2O_7^{2-}(aq) \rightarrow 2 Cr^{3+}(aq) + 7 H_2O(l)$, there is 1 electron involved in the oxidation of the ferrous ion to ferric, but there is a coefficient of 6 in the equation. Pure liquids do not appear in the Nernst equation.

440. (C) Looking at the half-cell potentials, you can see that fluorine has a greater positive voltage; therefore, fluorine will oxidize chlorine, not the reverse, eliminating answer A. The halogens are undergoing reduction and are not reducing agents, eliminating answer B, and since fluorine has a larger positive reduction half-cell potential, it would be easier to oxidize chlorine than fluorine. The large positive half-cell reduction potentials indicate that both halogens are good oxidizing agents.

441. (C) Reduction is the gain of electrons, eliminating answers B and D, and the oxidation state decreases, eliminating answer A.

442. (A) Reaction 1 involves the loss of electrons and is oxidation. Reaction 2 involves the gain of electrons and is reduction. This eliminates answers B and C. Oxidation occurs at the anode, and reduction occurs at the cathode, eliminating answer D.

443. (C) In the reaction, the Sn(IV) to Sn(II) reaction is reduction, and the reaction involving the mercury is oxidation. Therefore, you must reverse the mercury half-reaction and change the sign on the half-cell potential to -0.79 V. Adding the two half-cell potentials together to get the overall cell potential gives 0.15 V $+ (-0.79$ V$) = -0.64$ V. Since the cell potential is negative, the reaction must be nonspontaneous.

444. (B) The oxidizing agent will be the reactant that is being reduced in the reaction. The $Fe^{2+}(aq) \rightarrow Fe^{3+}(aq) + e^-$ is oxidation, so $Cr_2O_7^{2-}(aq)$ is being reduced and is the oxidizing agent.

445. (C) Since this will be a galvanic cell, the cell potential must be positive. One of the reactions must be reversed, changing the sign of the half-cell potential. Electron loss and gain will be equalized by the use of appropriate coefficients, but that will not affect the half-cell potentials. The only way that you can get an overall positive voltage is to reverse the second reaction and add the half-cell potentials: $E°_{cell} = +1.65 + 0.58 = +2.23$ V.

446. (A) To oxidize 1 mole of Sn(II) to Sn(IV) requires 2 moles of electrons; 2 moles of electrons are also required for the half-reaction involving the HClO. However, there was enough tin to require 3 moles of HClO, which means 6 moles of electrons were transferred. The ClO_3^- half-reaction involves 6 moles of electrons for every mole of ClO_3^-, so it would require 1 mole of ClO_3^-.

447. (D) The reaction is $Mg^{2+} + 2\ e^- \rightarrow Mg$. You can convert from g Mg $\rightarrow$ moles Mg $\rightarrow$ moles $e^- \rightarrow$ coulombs $\rightarrow$ current. The calculation (with rounding) is:

$$\frac{1.5\ g\ Mg}{1} \times \frac{1\ mol\ Mg}{24\ g} \times \frac{2\ mol\ e^-}{1\ mol\ Mg} \times \frac{100{,}000\ C}{1\ mol\ e^-} \times \frac{amp \cdot sec}{C} \times \frac{1}{6000\ sec}$$

≈ 2.0 amp

448. **(A)** The reaction would be $Ca^{2+} + 2\,e^- \rightarrow Ca$. The grams of Ca can be calculated (17 min = 1020 sec). The calculation (with rounding):

$$\frac{2.0\ \text{amp}}{1} \times \frac{1000\ \text{sec}}{1} \times \frac{1\ \text{C}}{\text{amp}\cdot\text{sec}} \times \frac{1\ \text{mol e}^-}{100{,}000\ \text{C}} \times \frac{1\ \text{mol Ca}}{2\ \text{mol e}^-} \times \frac{40\ \text{g}}{1\ \text{mol Ca}}$$

$$\approx 0.4\ \text{g Ca}$$

Choose the closest answer, answer A.

449. **(C)** At equilibrium, the cell potential will be zero, eliminating answer D. High pH will reduce the hydrogen ion concentrations, which will be in the denominator in the activity quotient, thus making the cell potential larger.

450. **(A)** The reaction is $Hg^{2+} + 2\,e^- \rightarrow Hg$. The grams of mercury produced can be calculated (with rounding):

$$\frac{2.0\ \text{amp}}{1} \times \frac{3600\ \text{sec}}{1\ \text{hr}} \times \frac{3\ \text{hr}}{1} \times \frac{1\ \text{C}}{1\ \text{amp}\cdot\text{sec}} \times \frac{1\ \text{mol e}^-}{100{,}000\ \text{C}} \times \frac{1\ \text{mol Hg}}{2\ \text{mol e}^-}$$

$$\times \frac{200\ \text{g Hg}}{1\ \text{mol}} \approx 22\ \text{g Hg}$$

Chapter 12: Final Review

451. **(C)** Barium is in the same family as calcium, so it behaves like Ca. Calcium is an important element/ion in many biological functions, including bone formation.

452. **(A)** The molar mass of $CuSO_4 \cdot 5H_2O$ would be the sum of the atomic masses: $[63.6 + 32.1 + (4 \times 16.00) + 5(2\,(1.0) + 16.00)]$ g/mol $\neq$ 250 g/mol.

453. **(C)** Only answers B and C contain halogens. Answer B: Cl_2O (chlorine in +1) $HClO_2$ (chlorine in +3); answer C: Cl_2O_5 (chlorine in +5) $HClO_3$ (chlorine in +5).

454. **(D)** The formula of calcium nitrate is $Ca(NO_3)_2$. In the 0.5 M $Ca(NO_3)_2$ solution, the $[NO_3^-]$ is 1.0 M (2:1 ratio between calcium ions and calcium nitrate). Then the solution is diluted: (0.100 L)(1.00 M) = (1.000 L)(? M); M = 0.10 M.

455. **(B)** The heat of vaporization depends on the mass and the strength of the intermolecular forces between the molecules. All of the compounds have approximately the same mass, so the determining factor would be the intermolecular forces. The compound in answer C, a hydrocarbon, would have only weak London forces as the intermolecular force. Answer D would have weak dipole-dipole attraction. Answer A has one hydrogen-bonding site, but answer B has two hydrogen-bonding sites, so its attractive force would be the largest and so would its heat of vaporization.

456. (A) The heat of vaporization depends upon the molar mass (C) and the intermolecular forces (answers B and D—these two are really the same). The heat of vaporization does not depend on the external pressure.

457. (B) All of the compounds are covalently bonded molecules except NaCl. NaCl is ionically bonded, and ionic compounds have higher melting points than the molecules with hydrogen bonding as the intermolecular force.

458. (D) $CHCl_3$ is covalently bonded—no ions produced. Answer B is a 1:2 salt, so it would produce 0.90 moles of ions; answer C is a 1:1 salt, so it would produce 0.90 moles of ions; answer D is a 1:3 salt and will produce 1.0 moles of ions.

459. (A) The IV fluid should be such that its osmotic pressure is close to that of blood plasma. The osmotic pressure of water is too low.

460. (B) In order to have hydrogen bonds, there must be a hydrogen atom attached to an O, N, or F. Answers A and C have $-OH$ groups that can hydrogen bond; answer D has an $-NH_2$ group, which can hydrogen bond. In answer B, the hydrogen atoms are bonded to the carbon.

461. (A) The lower the activation energy, the faster the reaction. Answer A has the lowest activation energy.

462. (C) Intermediates do not appear in the overall reaction as either reactants or products—they are both formed and consumed during the reaction. This is true of the chlorine and hydroxyl radicals.

463. (D) Adding $CHCl_3$ will not affect the equilibrium, since this is a covalently bonded molecule. Answer B is incorrect, because reducing the Pt concentration would force the reaction back to the left, reducing the cisplatin concentration. Answer C is incorrect, because reducing the ammonia concentration again shifts the equilibrium to the left. Answer D is correct, because increasing the ammonia concentration shifts the equilibrium to the right, increasing the cisplatin concentration.

464. (B) Lowering the pH increases the number of hydrogen ions that would be available to react with the weakly basic oxalate ion. This would increase the solubility of the calcium oxalate, since it is reacting with the acid (lower pH).

465. (A) If the reaction quotient of a solution exceeds the K_{sp}, then a precipitate will form. The system is at equilibrium if the reaction quotient equals the K_{sp}.

466. (C) Adding a catalyst affects both the forward and reverse reactions of the equilibrium system equally, so no net change in the equilibrium constant is observed. However, the system will get to equilibrium faster.

467. (D) This is a case of mixing thermodynamics and kinetics. The activation energy gives information about the kinetics of the reaction but gives no information about the energy change (thermodynamics).

468. (A) Catalysts do not change the heats of reaction; catalysts affect the kinetics (speed) of reaction.

469. (D) The heat of reaction depends on the difference in the energy levels of the reactants and products. It does not depend on the activation energy or transition state.

470. (A) If you calculate all the oxidation numbers for the elements in the reaction, you find that there are changes that identify both an oxidation and a reduction occurring. (You need to find only one oxidation or reduction, since if one is present, the other must be there.) Respiration is a redox process.

471. (B) In batteries, you have a difference of potential that is related to the half-reactions and the reactant concentrations; as the reaction proceeds, the concentrations decrease until there is no longer a potential difference—the cell is at equilibrium.

472. (A) In order for each electron transfer to be spontaneous, the voltage for each electron transfer must be positive. Starting with the largest voltage, you can form a redox reaction with that one and the one with the next highest voltage; so when that one is reversed and the sign on the voltage is made negative, the result is a positive total voltage. Then do that for the third reaction and the second reaction, and so on. The correct sequence goes from the one with the highest reduction potential to the one with the lowest, answer A. Changing the order will yield at least one negative sum (non-spontaneous) result.

473. (A) The theoretical yield is the maximum amount of a product that can be formed. Temperature is not normally considered; the amount of NaCl recovered is the actual yield. The concentration of the excess reactant does not matter, since that reactant is in excess. The maximum amount of reactant that can be formed is determined by the amount of the limiting reactant.

474. (A) Balancing the nuclear equation (making sure that you have the same sum of mass numbers and atomic numbers on both sides) gives:

$$^{235}_{92}U + ^{1}_{0}n \rightarrow ^{140}_{55}Cs + 3\,^{1}_{0}n + ^{93}_{37}Rb$$

Answer A is correct.

475. (A) Ge is in the IVA family, meaning that it has 4 valence electrons, 2 in the 4s and 2 in the 4p, with a filled 3d. There must be 32 total electrons. Answer A has this configuration.

476. (A) Answer A has (13 p + 14 n) versus (14 p + 13 n), answer B has (11 p + 12 n) versus (12 p + 12 n), answer C has (22 p + 26 n) versus (20 p + 28 n), and answer D has (15 p + 17 n) versus (17 p + 19 n). Answer A contains mirror nuclei.

477. (A) Sodium metal has 11 electrons, eliminating answers C and D, which show only 10 electrons. Sodium's single valence electron is in the 3s orbital, so answer B is the ground-state configuration for sodium. It would be that 3s electron that would gain energy and become excited to the closest higher energy level, the 3p.

478. (A) As you proceed from left to right in a period, the electrons are added to the same major energy level, but the increasing effective nuclear charge causes the atomic radius to decrease. This eliminates answers C and D. Electronegativity is the attraction an atom has for a bonding pair of electrons, so the smaller the atom, the greater the effective nuclear charge and the greater the electronegativity.

479. (D) HCl would be polar, since it is a linear molecule with the atoms having a difference in electronegativities. CH_3Cl would be polar, since the chlorine end of the molecule would pull electron density toward it. H_2S is a bent molecule, like water, with the sulfur being more electronegative than hydrogen. CF_4, even though the elements have a large difference in electronegativities, is a symmetrical molecule (tetrahedral like methane) and thus is nonpolar.

480. (B) Sp hybridization involves the blending of an s and a p orbital. The two resultant sp hybrid orbitals lie 180° apart. This is a linear orientation.

481. (B) These are all members of the same group, the alkaline earth metals. Valence electrons are more easily lost the farther they are from the nucleus. The atomic radius increases from top to bottom within a group. Therefore, the least reactive should be the group member closest to the top of the periodic table, in this case Be.

482. (A) The melting point of an ionic substance is related to the strength of the attractive force between the ions. Both the rubidium cation and the iodide anion are rather large ions, much larger than any of the other ions in the other choices. They are also monovalent ions, so the charge density on each is small. Therefore, the attractive force holding them together is weaker than for the other choices and the melting point is lower.

483. (C) The partial pressure of a gas may be calculated by multiplying the mole fraction of the gas by the total pressure. The total number of moles of gas is 8 moles, and the mole fraction of hydrogen is ⅛. Therefore, the partial pressure of hydrogen is ⅛ × 1,000 torr = 125 torr.

484. (A) The rate of effusion of a gas is inversely related to the square root of its molar mass (Graham's law). The molar mass of methane, CH_4, is 16 g/mol and is 32 g/mol for oxygen gas, O_2. The ratio of methane to oxygen effusion would be the square root of the molar mass of oxygen divided by the molar mass of methane. Rate of effusion of methane to oxygen = $(32 \text{ g/mol} / 16 \text{ g/mol})^{1/2} = 1.4$.

485. (A) The volume of a gas is directly proportional to the Kelvin temperature. The initial temperature would be $(227 + 273) \text{ K} = 500 \text{ K}$, and the final temperature would be $(727 + 273) \text{ K} = 1,000 \text{ K}$. Since the final temperature is larger than the initial, the gas will expand, so you must multiply the initial volume by a fraction greater than 1: $(1,000 \text{ K} / 500 \text{ K}) \times 500 \text{ cm}^3 = 1,000 \text{ cm}^3$.

486. (A) For a real gas to approach ideal behavior, the pressure must be low to minimize the effect of the gas's volume (and intermolecular forces of attraction), and the temperature should be high to minimize the attractive forces between gas particles.

487. (C) One mole of $Ce(NO_3)_3$ will furnish 4 moles of ions, 3 moles of nitrate anions, and 1 mole of cerium cations. If you want 1.0 mole of ions, you will need 0.25 moles of $Ce(NO_3)_3$.

488. (A) Answer B is unlikely because it involves a three-body collision. Answer C is unlikely because there is only 1 NO in the (slow) rate-determining step. Answer D is unlikely because there is no NO in the slow step. Answer A is quite likely because in the slow step, you can substitute $[Br_2][NO]$ for the $[NOBr_2]$, and that would give the observed rate law.

489. (A) By Le Chatelier's principle, if a system is at equilibrium and one of the concentrations is increased, the reaction shifts to the opposite side to replace it. The value of the equilibrium constant does not change. Therefore, the reaction will shift to the right.

490. (A) The best choice of a buffer is one whose pK_a is as close to the desired pH as possible. Then you can fine-tune the pH by adjusting the concentrations as shown by the Henderson-Hasselbalch equation: $pH = pK_a + \log([\text{base}]/[\text{acid}])$. HBrO has a K_a closest to the desired pH.

491. (C) Conjugate acid-base pairs differ by a single H^+. The conjugate bases have one less hydrogen ion. In answer A, the phosphate ion has 2 H^+ ions removed. In answer B, the charge on the NH_2 is incorrect. In answer D, only the conjugate acids are shown.

492. (C) $pK_a + pK_b = 14$; $pK_b = 14 - pK_a$. The $pK_a = -\log K_a$, so $pK_b = 14 - (-\log K_a)$, or $14 + pK_a$. Therefore, $pK_b = 14 + (\log 3.5 \times 10^{-4})$.

493. (A) With a $K_a = 5.1 \times 10^{-4}$, the pK_a should be between 3 and 4 bases upon the exponent; 3.3 is the only answer in that range.

494. (A) The entropy change for the reaction ($\Delta S°$) would be the (entropy of the products) − (entropy of the reactants): 1 mol(290 J/mol·K) − [1 mol(198 J/mol·K) + 1 mol(223 J/mol·K)] = −131 J/K.

495. (A) As the reaction proceeds, the amount of aluminum metal decreases, not increases as indicated by answer C. As the reaction proceeds, the concentration of hydrogen ions will decrease and the pH of the solution will increase (becoming more basic).

496. (C) The boiling point is a function of the external pressure, because a substance boils when its vapor pressure equals the external pressure. Decreasing the external pressure decreases the boiling point.

497. (A) Answer A is the only one that is a free radical (having an unpaired electron).

498. (D) Carbon dioxide, with its two double bonds, is linear (eliminating answers A and C). There would be a C=O bond, so there would be only three groups around the C, making the geometry trigonal planar.

499. (D) During a half-life, the amount you start with reduces by one-half. After the first half-life, you have 50% remaining; second half-life, 25%; third half-life, 12.5%; fourth half-life, 6.25%.

500. (D) Since you need only an approximate value of pK_{a1}, you can make the approximation that a K_a of $8 \times 10^{-3} \neq 1 \times 10^{-2}$. Therefore, the pK_{a1} should be approximately 2. Answer D, 2.1, is very close to this, much closer than any of the other answers.